The Journal of the International Space Elevator Consortium

(http://www.isec.org)

ACKNOWLEDGEMENTS

CLIMB – The Journal of Space Elevator developments and technology

Official Publication of the International Space Elevator Consortium (ISEC)
http://www.isec.org

Publication & Review Committee

Editor-in-Chief
　Ted Semon
　President and Director, ISEC

Chief Technical Editor
　Ben Shelef
　Director

Technical Review Committee
　Stephen Cohen, M. Eng.
　　Editor, The Engineer's Pulse
　Blaise Gassend, PhD
　　San Carlos, CA
　Benjamin H. Jarrell
　　Jarrell & Doty, P.C.
　Martin Lades, PhD
　　Nuremberg, Germany

ISEC Officers & Directors

Officers:
　President: Ted Semon
　Vice President: Peter Swan
　Secretary : Martin Lades
　Treasurer: Robert "Skip" Penny

Directors:
　David Horn
　Martin Lades, PhD
　Bryan Laubscher, PhD
　Robert "Skip" Penny
　Ted Semon
　Ben Shelef
　Peter Swan, PhD

CLIMB (Print version ISBN 978-1-304-24055-2, eVersion ISBN 978-0-9854262-4-8) is published by the International Space Elevator Consortium (ISEC), 709A N Shoreline Blvd, Mountain View, CA, 94043, USA. Please direct all enquiries to CLIMB@isec.org.

Editorial Communications should be sent to the International Space Elevator Consortium, 709A N Shoreline Blvd, Mountain View, CA. 94043, USA or emailed to CLIMB@isec.org.

Subscription Rates: The Print version of CLIMB is offered free, to all ISEC Members in good standing with membership level Professional or above. The electronic version of CLIMB is offered free to all ISEC Members in good standing. Individual copies may be purchased from the ISEC store (located on the ISEC website; http://www.isec.org). For questions or more information, please email us at CLIMB@isec.org.

MISSION: Our Mission statement reads *"ISEC promotes the development, construction and operation of a space elevator as a revolutionary and efficient way to space for all humanity."* ISEC works towards this goal by publicizing and promoting the concept, promoting and encouraging research in technologies necessary to build a space elevator and promoting and encouraging research into all aspects of space elevator technology.

Second Printing. Copyright © 2013 by the International Space Elevator Consortium. All rights reserved. No part of this work may be reproduced or translated in any way without permission from the copyright owner. Permission to reproduce all or part of this publication in any form must be obtained in writing from the President of ISEC. Requests for reprints should be sent to CLIMB@isec.org.

Disclaimer: The ideas and opinions expressed in CLIMB do not necessarily reflect those of ISEC or the Editors of CLIMB unless so stated. Articles contained in the Papers section of this journal have been reviewed and approved by the CLIMB Technical Committee but again; do not necessarily reflect those of ISEC or the Editors of CLIMB unless so stated.

Print version published by lulu.com. eVersion published by ISEC.

CONTENTS

INTRODUCTION

ACKNOWLEDGEMENTS	iii
CONTENTS	v
FOREWORD Pearson, J.	vii
PREFACE Semon, T	ix

PAPERS

GETTING THE MOST OUT OF NANOTUBES: GUIDANCE FROM FRACTURE PHYSICS AND ATOMISTIC SIMULATIONS Artyukhov, V. / Liu, Y. / Yakobson, B.	3
THE EFFECT OF COLLAPSED NANOTUBES ON NANOTUBE BUNDLE STRENGTH Pugno, N.	11
SATELLITE PLACEMENT USING THE SPACE ELEVATOR Cohen, S. / Misra, A.K.	17
SPACE ELEVATOR DEPLOYMENT Dempsey, J.	27
TRANSVERSE VIBRATION OF THE SPACE ELEVATOR TETHER WITH SPACEPORTS Ambartsumian, S.A. / Belubekyan, M.V. / Ghazaryan, K.B. / Ghazaryan, R. A	53
SPACE ELEVATOR POWER SYSTEM ANALYSIS AND OPTIMIZATION Shelef, B.	65

OTHER READING

THE REAL HISTORY OF THE SPACE ELEVATOR Pearson, J.	75
SPACE ELEVATOR INITIAL CONSTRUCTION MISSION OVERVIEW Lang, D.	85
LUNAR ANCHORED SATELLITE TEST Pearson, J.	105
ASTEROID SLINGSHOT EXPRESS - TETHER-BASED SAMPLE RETURN Shelef, B.	113
ISEC THEME POSTERS Chase, F	119

Jerome Pearson

FOREWORD

Since Yuri Artsutanov and I met again at the Space Elevator Conference in 2010, there has been continued progress on realizing this dream of a connection between Earth and sky. There is new hope that the three problems of material strength, radiation, and collisions can be overcome. The Space Elevator Conference of 2010 addressed space debris and reducing the collision risks, and the 2011 Conference featured researchers and progress in nanotubes to meet the requirements for high strength materials and the 2012 Conference furthered discussions of how a Space Elevator could be operated. The Space Elevator Conference of 2013 presents new ideas and concepts to bring the space elevator closer to reality.

I still remember my initial despair at the phenomenal strength of materials that would be required for the classical Earth space elevator balanced about the geostationary altitude that Yuri Artsutanov and I invented. Georg von Tiesenhausen, one of Von Braun's Project Paper Clip engineers from Peenemunde who was then at NASA Marshall, suggested that I look at a lunar space elevator, which could be built using existing composite materials. I did this and published in 1979, followed shortly by Yuri's publication. (We later discovered that the Russian Friedrich Tsander had discussed a lunar space elevator in his papers of 1929. It seems there is truly "nothing new under the sun"!)

The combination of the Earth space elevator and the lunar space elevator, proposed by the LiftPort Group, would allow a truly revolutionary Earth-Moon transportation system, and lead to the development of cis-lunar space. This would also allow the large-scale use of lunar regolith for radiation shielding, and for construction of unprecedented space structures in Earth orbit.

Finally, the combination of the Lofstrom Loop as the base of the classical space elevator, as described at the 2012 Conference by John Knapman, also allows for some exciting possibilities. And although Paul Birch died in 2012, I still look forward to the possibility of realizing his orbital ring approach to a low-altitude space elevator, which allows other exciting possibilities.

Each new Space Elevator Conference brings new ideas and solid progress, and brings us closer to realizing our dreams. We are all indebted to the efforts of ISEC for making these conferences possible, and for keeping attention focused on the space elevator at the annual International Astronautical Congress. This Volume 2 of C^LIMB is an appropriate supplement to the 2013 Space Elevator Conference, and I look forward to being part of a successful and memorable conference.

Jerome Pearson

July 2013

Mount Pleasant, South Carolina

PREFACE

Welcome to the second issue of C^LIMB, a Journal devoted solely to the space elevator. Producing such a Journal is one of the major 'products' of the International Space Elevator Consortium (ISEC). C^LIMB presents some of the best, peer-reviewed articles written on space elevator-related topics in the past several months as well as some additional papers we believe will be of interest to our readers. ISEC plans on producing additional issues of C^LIMB on a regular basis in the future.

During the past 12 months, major developments in the space elevator community include:

- Multiple Space elevator competitions were held in both Europe and Japan.
- Conferences devoted wholly or partially to the Space Elevator now occur on a regular basis in the United States, Japan and Europe. At the 2011 US Space Elevator Conference, ISEC sponsored the attendance of several leading academic researchers specializing in carbon nanotubes (CNTs). These researchers gave an excellent overview on exciting new developments in this field. And in 2012, for the first time, ISEC organized the annual American Space Elevator Conference and will continue to do so going forward.
- ISEC has recently released its second position paper, this on *Operating and Maintaining a Space Elevator* (the ISEC theme for 2012). The first draft of this paper was presented at the 2012 Space Elevator Conference with the final report issued in the Spring of 2013. ISEC plans on releasing studies like this every year.
- The International Academy of Astronautics has completed the review of a four year, 41 authors study and will submit it to the publisher. The title is *Space Elevators: An Assessment of the Technological Feasibility and the Way Forward*. The Space Elevator Study Group addressed its feasibility and proposed roadmaps. Most of the group are academy elected members; and, others are technical experts invited to participate in the study. Some of the key players in this group are ISEC Directors.
- It now appears that there may be an alternative to Carbon Nanotubes and that is Boron Nitride Nanotubes (BNNTs). While not naturally occurring in nature, this substance promises nearly all of the strength of CNTs and may be much more amenable to manufacturing processes than CNTs are. Options are a good thing!
- And, a development not directed at the Space Elevator community, but clearly of interest to it was the fourth annual Carbon Nanotube conference sponsored by and held at the University of Cincinnati. In previous conferences, the emphasis was on how carbon nanotubes could be used in electrical devices, as artificial muscles, mixed with ceramics, etc. However, in the 2011 and 2012 conference, fully half of the presentations at least touched on maximizing the specific strength of CNT threads and some papers were devoted solely to this topic. It will take this kind of concentrated interest by the right parties to lead to the breakthrough we're all waiting for.

Let me close by encouraging you to consider joining ISEC. Our Mission Statement, *"ISEC promotes the development, construction and operation of a space elevator as a revolutionary and efficient way to space for all humanity"* says it all. We are doing everything we can to promote the idea of a space elevator and we need your help and your membership funds to continue to move forward. Please visit our website; http://www.isec.org to learn more about how you can become involved in this magnificent project, and don't forget to sign up for our free newsletter, keeping you up-to-date on space elevator- related developments.

Ted Semon, President – The International Space Elevator Consortium

PAPERS

GETTING THE MOST OUT OF NANOTUBES: GUIDANCE FROM FRACTURE PHYSICS AND ATOMISTIC SIMULATIONS

Vasilii I. Artyukhov (artyukhov@rice.edu) and Yuanyue Liu
Department of Mechanical Engineering and Materials Science, Rice University, Houston, 77025 TX

Boris I. Yakobson (biy@rice.edu)
Department of Chemistry, Department of Mechanical Engineering and Materials Science, and Smalley Institute for Nanoscale Science and Technology, Rice University, Houston, 77025 TX

Abstract: Fracture of graphene – the fundamental crystal lattice underlying all sp^2 carbon nanomaterials – is studied theoretically basing on an expression for the energy of graphene edge as a function of edge orientation. The form of this dependence results in an unusual behavior: cracks in graphene are predicted to follow only straight paths along the two fundamental directions of graphene lattice. Molecular dynamics simulations of graphene sheet fracture support this conclusion. This leads to the proposal of a novel fracture mechanism for large-diameter chiral carbon nanotubes, where cracks are predicted to follow a helical path along the tube circumference. Again, atomistic simulations add evidence supporting this exotic mechanism. These findings provide important guidelines for the maximization of strength and toughness of graphitic carbon nanomaterials, and suggest possible benefits of using boron nitride-based nanomaterials.

Introduction

Non-rocket means of space launch are attractive because of the projected lower costs, as compared to the usual means of getting to space using rockets. In fact, non-rocket space launch in some sense predates rocket technology: one example is the space cannon from Jules Vernes' 1865 *From the Earth to the Moon*. Another is the space tower proposed by Tsiolkovskii in 1895 (in *Speculations about Earth and Sky and on Vesta*), eight years before he published his ideal rocket equation.

Generally, most non-rocket space launch proposals can be roughly categorized into the following four classes:

Projectile: space cannon, railgun. Designs from this category may be the simplest and cheapest possible, but the extreme accelerations involved are unsuitable for delicate equipment or human spaceflight. An extremely cheap and effective approach would be to utilize the energy of an underground/underwater nuclear blast for payload acceleration [1], but this creates obvious political problems.

Compressive: space towers. The original idea appears impractical as the required base cross-section area (and the material cost of the structure) scales exponentially with height, and the scale height for currently available materials prohibits sufficiently tall towers. One proposal discusses a pneumatic tower claiming achievable height of 200 km, which still seems insufficient, so compressive structures presently can only be viewed as possible complements to reduce costs for other approaches.

Tensile: skyhook, bolus, geostationary space elevator. Tensile failure has fewer diverse modes than compressive, and the necessary specific strength sufficient to sustain a full geostationary structure appears achievable with known materials, specifically, carbon nanotubes or other forms of graphitic (with atoms in the sp^2 hybridization state) carbon. However, even these advanced materials leave only very little safety margin (see below), and reaching the necessary performance in terms of specific strength calls for not just incremental improvements of existing carbon fiber materials, but also for a very good and detailed theoretical understanding of mechanical failure of such materials at the fundamental level.

Dynamic: several designs have been put forward where inertia of a rotating structure (orbital ring [2]) or a moving payload (space fountain, launch loop) would counteract Earth's gravity, relaxing the requirements on material strength. However, such designs typically rely on vacuum and magnetic levitation to minimize friction in moving parts, which raises material costs and probably depends on development of radically new materials, e.g., superconductors.

In summary, it appears that only tensile structures in this classification look achievable without any radical scientific breakthrough. Here we focus on the geostationary space elevator [3-8] design, as being seemingly the most mature of all the non-rocket space launch concepts.

Although a tapered design permits an arbitrarily long tether made from any material, the use of softer materials (in terms of specific strength as measured in units of Pa/(g/cm^3), or Yuri) would result in rapidly increasing cable mass [4]. The requirement that the elevator must be able to lift at least its own mass of material per ~1 year for maintenance needs results in a trade-off between the specific strength of material and climber power. Current estimates place the feasibility range at around 25–30 MYuri [9]. Given a tensile breaking strength of 100–120 GPa and a density of 2 g/cm^3, the back-of-the-envelope estimate of specific strength of graphitic materials is around 50–60 MYuri, or roughly twice the feasibility threshold. (It is also of prime importance that sp^2 carbon behaves as a brittle material at relevant temperatures [10], which rules out plastic failure.)

Thus, graphitic carbon (in the form of carbon nanotubes or nanofibers) seems an attractive candidate that may be actually fit for the task. However, the margin of safety of the above result is uncomfortably low. In this context, it is of prime importance to pursue a detailed understanding of fracture of graphitic carbon materials at the fundamental atomistic level.

Basic theory: direction-dependent strength of graphene

The ultimate carbon fibers, carbon nanotubes, can be thought of as a graphene layer rolled up into a cylinder. In this respect it is very instructive to study the strength of graphene lattice in order to establish the similarities and differences between the flat and rolled-up cases.

From the continuum mechanics viewpoint, crack propagation during the fracture of a material is governed by so-called critical stress intensity factor, $K_C \sim (\gamma Y)^{1/2}$, where Y is the Young's modulus of the material, and γ is the surface energy. In two dimensions, Y is replaced by the 2D in-plane stiffness, which is isotropic for graphene, whereas γ stands for the 1D edge energy. The latter quantity is direction-dependent, and this dependence has a simple analytical form [11]:

$$\gamma(\chi) = 2\gamma_A \sin(\chi) + 2\gamma_Z \sin(30°-\chi) = |\gamma|\cos(\chi+C) \qquad (1)$$

Here, χ is the angle between the edge and the zigzag direction of the graphene lattice, γ_A and γ_Z are the edge energies of perfect zigzag and armchair edges. (Strictly speaking, Eq. 1 has to be augmented with a so-called *AZ*-mix energy correction [11] that is small but of certain fundamental importance: $\delta \cdot 4\sin(\chi)\sin(30°-\chi)/\cos(30°+\chi)$, where δ is typically on the order of $-0.01|\gamma|$.)

The Wulff construction [12,13] for the equilibrium shape of a graphene crystal using Eq. 1 predicts that both zigzag and armchair edges can be stable when $(\sqrt{3})/2 < \gamma_Z/\gamma_A < 2/(\sqrt{3})$, whereas outside this range, only one edge type (with lower energy γ) will be observed. This "bi-stable" range corresponds to a concave shape of $\gamma(\chi)$, i.e. such that it has a maximum at a certain value of χ. Normally, for graphene in vacuum, the more reliable *ab initio* (from the first principles of quantum mechanics) calculations predict that this is not the case and zigzag edge is unstable, whereas with some popular empirical interatomic potentials the situation is reversed [11], and only zigzag edges are predicted to be stable (see, e.g., [14]). While it is obvious that only one of these situations is correct in the literal sense, i.e., for an actual clean graphene edge in vacuum, we have to be careful because subtle differences in the environment can affect the γ_Z/γ_A ratio; hence, we can use different models that predict different γ_Z/γ_A ratios as useful proxies for situations other than vacuum, even if in the literal sense the result may not appear physically relevant. We must also point out that the zigzag edge can assume a reconstructed form for which edge energy almost equals γ_A [15], which would result in *ab initio* calculations predicting a concave shape of $\gamma(\chi)$ even in vacuum. Energy barrier estimates for such a reconstruction suggest that it should not be expected to occur at moderate temperatures [16], but mechanical fracture is associated with rapid release of large amounts of elastic energy, which could facilitate the reconstruction. This mechanism will be explored in more details in an upcoming study, but for the present purposes the bottom line is that, if Eq. 1 holds, different atomistic models can be used to simulate fracture of graphene-based materials in varying conditions.

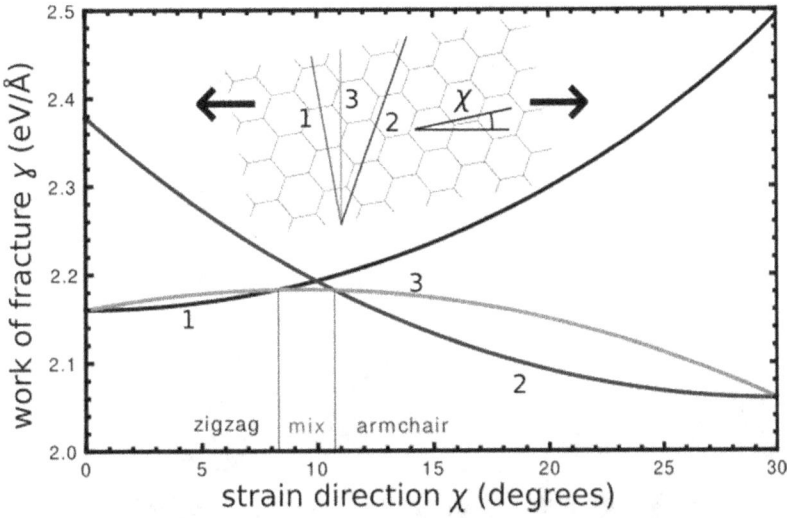

Figure 1. Energy of graphene fracture as a function of direction of applied tension for different crack trajectories: (1) along the zigzag direction, (2) along the armchair direction, (3) normal to the load (intermediate direction, mixed-type edge), and (4) normal to the load but without the AZ-mix correction taken into account. Only either pure armchair or pure zigzag direction is stable for all orientations χ except for a narrow window determined by the AZ-mix correction magnitude.

How does Eq. 1 determine the direction dependence of graphene strength? Fig, 1 shows the plot of direction-dependent crack energies for cracks taking different paths. Graphene sheet is loaded at an angle χ to the armchair direction of graphene, and the different curves correspond to different modes of crack propagation: the concave lines plot Eq. 1 without and with the AZ-mix energy correction, corresponding to a crack normal to the load direction, whereas the two convex lines represent straight cracks along the armchair and zigzag direction. For any chosen load orientation χ, the lowest energy line determines the path the crack will take, i.e., that with lowest K_C. It can be seen that Eq. 1 predicts straight cracks along either armchair or zigzag direction for almost any χ value except for a narrow window (determined by the magnitude of the AZ-mix correction). The particular edge energy values γ_Z and γ_A used for plotting Fig. 1 were calculated using the recently developed ReaxFF reactive forcefield [17,18] as implemented in the LAMMPS code [19].

Molecular dynamics simulations of graphene fracture

In order to confirm the applicability of the above analytical treatment, we carried out a series of molecular dynamics simulations of graphene sheets under load. For the first batch of runs, we started with approximately square-shaped flakes (Fig. 2, top) at three different orientations, $\chi = 0°, 13°, 30°$, with a notch on the left to initiate crack propagation. The sheets were pulled at a constant rate of ~10 m/s which is sufficiently small compared to sound velocity (~20 km/s), so that the loading was quasistatic. The bottom of Fig. 2 shows the results of the runs. In the left two panels (d, e), straight tears are seen that are perfectly aligned with the armchair or zigzag lattice directions. The rightmost panel (f) shows a crack that starts taking the shortest path normal to the load, which is what would be expected of a usual isotropic material, but then makes a sudden sharp turn and follows a strikingly straight path along the armchair direction. (Redirection of tears in graphene was observed in simulations before [20], but no discussion of the underlying reasons had been provided.)

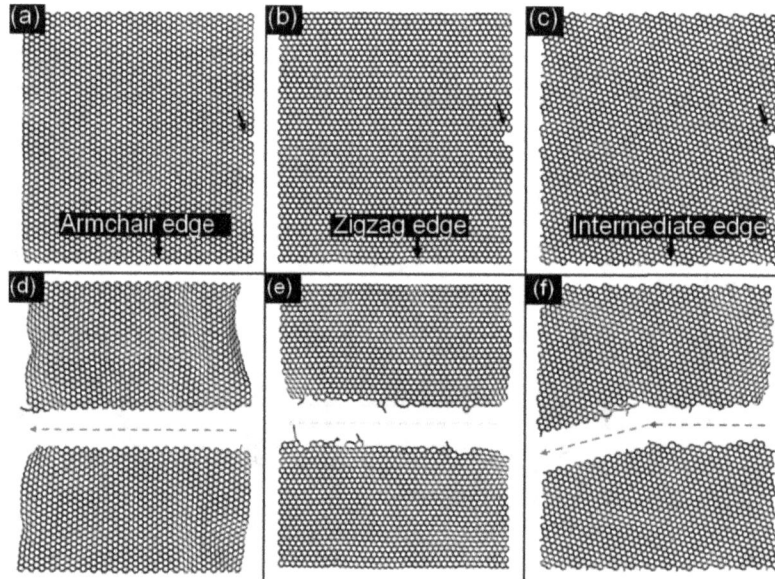

Figure 2. Molecular dynamics simulations of fracture of graphene sheets under different orientations of applied strain ($\chi = 30°, 0°, 13°$ from left to right): (a-c) starting configurations, (d-f) final configurations. Strain is applied at a constant rate.

In the second series of molecular dynamic runs, we subjected the sheets to a much more complex stress pattern. Specifically, square-shaped graphene sheets were pulled along two corners, as highlighted in the left panel of Fig. 3 (a). The right part of Fig. 3 (b-e) shows snapshots from four runs (χ = 0°, 13°, 17°, 30°) at the same simulated loading duration. (The loading was performed not at a constant rate but with a linearly increasing external force instead, and it was more rapid than in the above cases.) It can be seen that, whereas the particular details differ between the orientations, the main features remain consistent: tears invariably follow almost perfectly along either armchair or zigzag directions with sharp turns and almost no intermediate-direction segments.

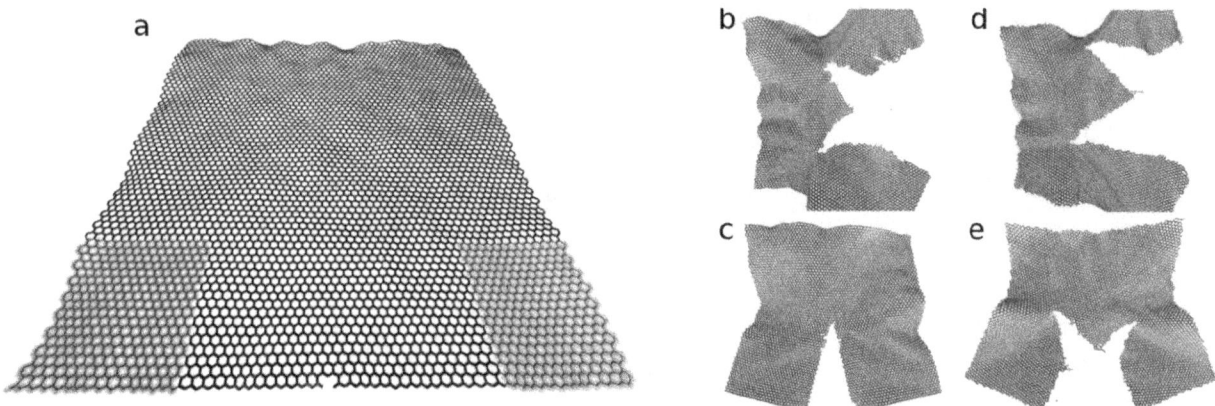

Figure 3. Molecular dynamics simulations of graphene fracture under a complex loading pattern: (a) graphene sheets are pulled by two corners as highlighted. The right part shows snapshots for the different lattice orientations χ = (b) 0°, (c) 30°, (d) 13°, (e) 17°.

The analytical calculations and molecular dynamics results presented above allowed us to provide explanation for experimental observations of ripping of graphene, where tears were seen to follow peculiar trajectories composed of straight line segments making sharp turns, but always aligned with zigzag or armchair direction of graphene, with no detectable presence of intermediate-orientation segments [21].

Implications for carbon nanotube strength

The above discussion applies directly to carbon nanotubes, but now χ denotes the chiral angle of the nanotube. In this case, line 3 in Fig. 1 corresponds to the shortest possible tear along the circumference of the nanotube. What do the other two crack trajectories in the figure correspond to for nanotubes? Unless χ = 0 or 30° (zigzag or armchair type nanotubes, respectively), a crack going purely along either of the two fundamental directions of the nanotube would take a helical path around the tube axis, effectively "unzipping" the nanotube into a graphene nanoribbon. If this were literally true, the work required to break a nanotube would scale with its length, i.e., its toughness would increase effectively indefinitely.

A series of molecular dynamics simulations were carried out for single-walled carbon nanotubes under constant-rate tensile load with chiral angles falling into the left (higher-energy) part of Fig. 2, χ = 5°, 10°, and 15°, and increasing gradually the tube diameter d = 1, 2, 4 nm. Fig. 4 shows a sequence of snapshots from the simulation of a (42,14) nanotube ($\chi = 10°$, $d = 4$ nm). Upon reaching a critical elongation, the pre-notched nanotube develops a crack that begins to follow a helical path. As the crack propagates to the opposite side of the tube and wraps around, it makes a turn with the two crack tips "homing in" onto each other.[1]

In the particular simulation shown in Fig. 4, the crack didn't actually close to form a loop, and an indication of the onset of spiral unzipping is seen. However, for all other 8 nanotubes that were simulated, the final configurations were completely disconnected in the middle. This is, of course, not surprising. As the crack makes its way around the tube, its two tips come within a distance of each other on the order of $\sim \pi d \sin(30°-\chi)$, assuming crack propagates along the armchair direction. Unless d is large, this results in self-interaction that causes significant redistribution of local strain [22], and the crack tips experience effective attraction enough to redirect them toward each other and close the crack into a loop separating the nanotube in two halves.

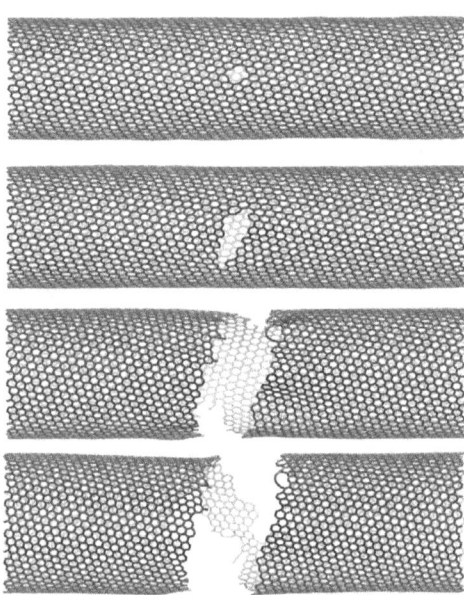

Figure 4. Sequential snapshots from a molecular dynamics simulation of a (42,14) nanotube ($\chi = 10°$, $d = 4$ nm) under tensile load, showing the onset of helical "unzipping".

Conclusions

In the present work, we have outlined a theory of fracture of graphene based on the analytical expression for direction-dependent edge energy. The theory predicts that cracks in graphene should follow mostly straight paths along the two fundamental directions in graphene lattice (armchair and zigzag). Molecular dynamics simulations of graphene sheets under tensile load support this analytical conclusion. Extension of the theory to carbon nanotubes has led to the prediction of a new mode of tensile failure for chiral tubes, with cracks following a helical path along the tube circumference. Whereas this mechanism can be expected to be relevant only in the "graphene-like" limit of large diameter tubes, molecular dynamics simulations provide evidence supporting it already for a 4 nm diameter tube, showing the onset of helical unzipping.

---

1 See http://www.owlnet.rice.edu/~va9/fracture-vids/spiral.mpg

In summary, this theory allows us to put forward the following recommendations to the builders of space elevators:

- Use nanotubes with specific chirality so as to maximize the edge energy $\gamma(\chi)$. This can lead to a sizable (up to 15%) gain in strength.
- Use large-diameter tubes. This can potentially lead to a large gain in toughness by directing the crack into a helical path and increasing its length.
- Maintain a well-controlled environment. Even minor presence of agents that can reduce the edge energy [11,23] can therefore drastically reduce the strength of the material [24].

Although the importance of developing means to produce carbon nanotubes with precisely controlled diameter and chirality has long since been recognized in the nanoelectronics community, where pure metallic or semiconductor nanotubes are needed to reliably fabricate electronics components, the dependence of mechanical properties of carbon nanotubes on chirality has received much less attention. The present contribution complements earlier work [10] in highlighting the importance of nanotube chirality control for the mechanical applications of carbon nanotubes.

On a last note, the results of this work shed a new light on another close relative of carbon nanotubes, namely, nanotubes of boron nitride (BN). Whereas they are generally perceived as less interesting than their carbon counterparts, edge energetics of BN is much more rich and tunable than that of graphene due to the binary chemical composition [25]. The mechanical stiffness of BN is almost as high as that of graphene [26-28], and its density exceeds graphene's by mere 2%. Finally, BN is more resistant to radiation damage than graphene-based materials are [29], and this is a major factor for long-term operation outside the protection of Earth's magnetosphere and especially within the van Allen belt [30]. We can thus envisage the advantages of the flexibility of BN edge energetics for finely engineering the performance of BN nanotubes in future construction materials.

References

[1] B. Wang, "The Nuclear Orion Home Run Shot, All Fallout Contained," *Next Big Future*. [Online]. Available: http://nextbigfuture.com/2009/02/nuclear-orion-home-run-shot-all-fallout.html. [Accessed: 28-Jul-2011].

[2] A. E. Yunitskii, "General Planetary Transport System," *Tekhnika Molodezhi (Technology for the Young)*, no. 6, pp. 34-36, 1982.

[3] Y. Artsutanov, "V kosmos na elektrovoze (To the Cosmos by Electric Train)," *Komsomolskaya Pravda*, 31-Jul-1960.

[4] J. D. Isaacs, A. C. Vine, H. Bradner, and G. E. Bachus, "Satellite Elongation into a True 'Sky-Hook'," *Science*, vol. 151, pp. 682-683, 1966.

[5] J. H. Shea, J. D. Isaacs, H. Bradner, G. E. Backus, and A. C. Vine, "Sky-Hook," *Science*, vol. 152, p. 800, 1966.

[6] V. Lvov, J. D. Issacs, H. Bradner, G. E. Backus, and A. C. Vine, "Sky-Hook: Old Idea," *Science*, vol. 158, pp. 946 -947, 1967.

[7] J. Pearson, "The orbital tower: A spacecraft launcher using the Earth's rotational energy," *Acta Astronautica*, vol. 2, pp. 785-799, 1975.

[8] B. C. Edwards, "Design and deployment of a space elevator," *Acta Astronautica*, vol. 47, pp. 735-744, 2000.

[9] "The Space Elevator Feasibility Condition," *The Spaceward Foundation*, 2008. [Online]. Available: http://www.spaceward.org/elevator-feasibility. [Accessed: 27-Jul-2011].

[10] T. Dumitrica, M. Hua, and B. Yakobson, "Symmetry-, time-, and temperature-dependent strength of carbon nanotubes," *Proc. Natl. Acad. Sci.*, vol. 103, pp. 6105-6109, 2006.

[11] Y. Liu, A. Dobrinsky, and B. Yakobson, "Graphene Edge from Armchair to Zigzag: The Origins of Nanotube Chirality?," *Phys. Rev. Lett.*, vol. 105, p. 235502, 2010.

[12] C. Herring, "Some Theorems on the Free Energies of Crystal Surfaces," *Phys. Rev.*, vol. 82, pp. 87-93, 1951.

[13] C. K. Gan and D. J. Srolovitz, "First-principles study of graphene edge properties and flake shapes," *Phys. Rev. B*, vol. 81, p. 125445, 2010.

[14] A. Omeltchenko, J. Yu, R. K. Kalia, and P. Vashishta, "Crack Front Propagation and Fracture in a Graphite Sheet: A Molecular-Dynamics Study on Parallel Computers," *Phys. Rev. Lett.*, vol. 78, p. 2148, 1997.

[15] P. Koskinen, S. Malola, and H. Häkkinen, "Self-Passivating Edge Reconstructions of Graphene," *Phys. Rev. Lett.*, vol. 101, p. 115502, 2008.

[16] J. M. H. Kroes, M. A. Akhukov, J. H. Los, N. Pineau, and A. Fasolino, "Mechanism and free-energy barrier of the type-57 reconstruction of the zigzag edge of graphene," *Phys. Rev. B*, vol. 83, p. 165411, 2011.

[17] A. C. T. van Duin, S. Dasgupta, F. Lorant, and W. A. Goddard, "ReaxFF: A Reactive Force Field for Hydrocarbons," *J. Phys. Chem. A*, vol. 105, pp. 9396-9409, 2001.

[18] J. E. Mueller, A. C. T. van Duin, and W. A. Goddard, "Development and Validation of ReaxFF Reactive Force Field for Hydrocarbon Chemistry Catalyzed by Nickel," *J. Phys. Chem. C*, vol. 114, pp. 4939-4949, 2010.

[19] S. Plimpton, "Fast Parallel Algorithms for Short-Range Molecular Dynamics," *J. Comput. Phys.*, vol. 117, pp. 1-19, 1995.

[20] T. Kawai, S. Okada, Y. Miyamoto, and H. Hiura, "Self-redirection of tearing edges in graphene: Tight-binding molecular dynamics simulations," *Phys. Rev. B*, vol. 80, p. 033401, 2009.

[21] K. Kim, V. I. Artyukhov, W. Regan, Y. Liu, M. F. Crommie, B. I. Yakobson, and A. Zettl, "Ripping Graphene: Preferred Directions," *Nano Lett.*, vol. 12, pp. 293–297, 2012.

[22] E. Bayart, A. Boudaoud, and M. Adda-Bedia, "Finite-Distance Singularities in the Tearing of Thin Sheets," *Phys. Rev. Lett.*, vol. 106, p. 194301, 2011.

[23] T. Wassmann, A. P. Seitsonen, A. M. Saitta, M. Lazzeri, and F. Mauri, "Structure, Stability, Edge States, and Aromaticity of Graphene Ribbons," *Phys. Rev. Lett.*, vol. 101, p. 096402, 2008.

[24] A. A. Griffith, "The Phenomena of Rupture and Flow in Solids," *Phil. Trans. R. Soc. London Ser. A,* vol. 221, pp. 163-198, 1921.

[25] Y. Liu, S. Bhowmick, and B. I. Yakobson, "BN White Graphene with 'Colorful' Edges: The Energies and Morphology," *Nano Lett.*, vol. 11, pp. 3113-3116, 2011.

[26] N. G. Chopra and A. Zettl, "Measurement of the elastic modulus of a multi-wall boron nitride nanotube," *Solid State Commun.*, vol. 105, pp. 297-300, 1998.

[27] E. Hernández, C. Goze, P. Bernier, and A. Rubio, "Elastic Properties of C and BxCyNz Composite Nanotubes," *Phys. Rev. Lett.*, vol. 80, p. 4502, 1998.

[28] K. N. Kudin, G. E. Scuseria, and B. I. Yakobson, "C2F, BN, and C nanoshell elasticity from ab initio computations," *Phys. Rev. B*, vol. 64, no. 23, p. 235406, 2001.

[29] J. S. Kim, K. B. Borisenko, V. Nicolosi, and A. I. Kirkland, "Controlled Radiation Damage and Edge Structures in Boron Nitride Membranes," *ACS Nano*, vol. 5, no. 5, pp. 3977-3986, 2011.

[30] A. M. Jorgensen, S. E. Patamia, and B. Gassend, "Passive radiation shielding considerations for the proposed space elevator," *Acta Astronautica*, vol. 60, no. 3, pp. 198-209, 2007.

THE EFFECT OF COLLAPSED NANOTUBES ON NANOTUBE BUNDLE STRENGTH

Nicola M. Pugno

1) Laboratory of Bio-Inspired & Graphene Nanomechanics, Department of Civil, Environmental and Mechanical Engineering, Università di Trento, via Mesiano, 77, I-38123 Trento, Italy,

2) Center for Materials and Microsystems, Fondazione Bruno Kessler,

Via Sommarive 18, I-38123 Povo (Trento), Italy.

nicola.pugno@unitn.it

Abstract: In this paper, we have evaluated the strength of a nanotube bundle, with or without collapsed nanotubes. The self-collapse can increase the strength up to a value of about 30%, suggesting a design towards Artsutanov's dream of the space elevator, thanks to the design of a 30MYuri strong tether. Graphene bundles are expected to be even stronger.

Keywords: nanotubes, bundles, self-collapse, space elevator, Artsutanov, 30MYuri.

Introduction

An explosion of interest in the scaling-up of buckypapers, nanotube bundles and graphene sheets is taking place in contemporary material science. In particular, nanostructures can be assembled (or well dispersed in a matrix) in order to produce new strong materials and structures. Recently, macroscopic buckypapers [1-5], nanotube bundles [5-12] and graphene sheets [13-16] have been realized. In spite of these fascinating achievements of the contemporary material science and chemistry we are evidently far from an optimal result. The reported mechanical strength of buckypapers and graphene sheets, for example, are comparable to that of a classical sheet of paper and macroscopic nanotube bundles have a strength still comparable to that of steel.

This paper, following [1], aims to extend the previous calculations performed by the same author, on the strength of nanotubes [17-20] or nanotube bundles [21-23] and assuming the intrinsic fracture of the composing nanotubes (i), for nanotube sliding (ii). For such a case, we have for the first time analytically calculated that single walled nanotubes with diameters larger than ~3nm will self-collapse in the bundle as a consequence of the van der Waals adhesion forces and that the self-collapse can enlarge the cable strength up to ~30%. This suggests the design of self-collapsed super-strong nanotube bundles, corresponding to a maximum cable strength of ~48GPa, comparable to the thermodynamic limit assuming intrinsic nanotube fracture of km-long cable (see [23], highlighted by Nature 450, 6, 2007). This result suggests that nanotube bundles are stronger than classical nanotube bundles. Such self-collapsed nanotube super-strong bundles are thus ideal for space elevator missions, where high strength is needed to prevent cable and mission failure. Note that the collapse under pressure, and even under atmospheric pressure, i.e. the self-collapse of nanotubes in bundle, was firstly investigated by atomistic simulations in [24]. Moreover, the self-collapse of nanotubes in a bundle has been recently experimentally observed [25]. Thus such super-strong bundles are becoming feasible. Graphene bundles are expected to be even stronger [26].

Self-buckling

The buckling pressure of a nanotube in a bundle can be calculated with the classical elastic buckling formula but including the "Laplace-like" surface adhesion pressure term [1]:

$$p_C = \frac{3N^\alpha D}{R^3} - \frac{\gamma}{R} \qquad (1)$$

where D is the graphene bending rigidity, N is the nanotube wall numbers, R is the nanotube external radius and γ is the surface energy. The first term in eq. (1), for $\alpha = 3$ is that governing the buckling of a perfectly elastic cylindrical long thin shell, whereas $\alpha = 1$ would describe fully independent walls.

From eq. (1) we derive the following condition for the self-collapse, i.e. collapse under zero pressure, of a nanotube in a bundle:

$$R \geq R_C^{(N)} = \sqrt{\frac{3N^\alpha D}{\gamma}} = \sqrt{6} R_0^{(N)} \qquad (2)$$

Taking $D = 0.11 \text{nN} \cdot \text{nm}$ and $\gamma = 0.18 \text{N/m}$ we find $2R_C^{(1)} \approx 2.7 \text{nm}$. Considering an intermediate coupling between the walls ($\alpha \approx 2$), the critical diameters for double and triple walled nanotubes are $2R_C^{(2)} \approx 5.4 \text{nm}$ and $2R_C^{(3)} \approx 8.1 \text{nm}$.

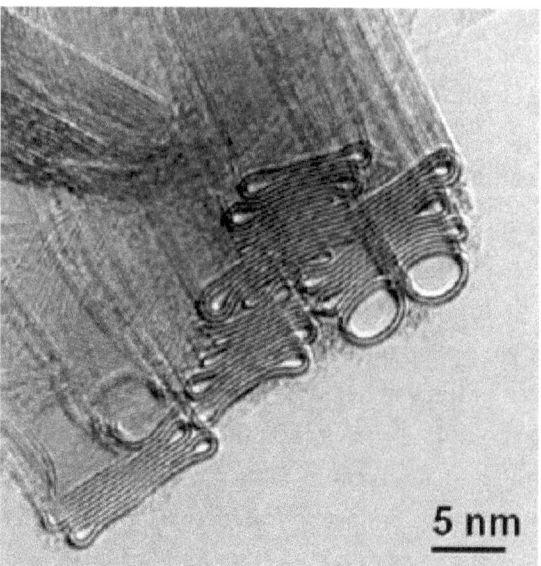

Figure 1: Self-collapsed nanotubes in a bundle [25].

In [25], 17 experimental observations on the self-collapse of nanotubes in a bundle have been reported, see Figure 1 and related Table 1. A number of 5 single walled nanotubes with diameters in the range 4.6-5.7nm were all observed as collapsed; moreover, while the 3 double walled nanotubes observed with internal diameters in the range 4.2-4.7nm (the effective diameters are larger by a factor of ~0.34/2nm) had not collapsed, the observed 8 double walled nanotubes with internal diameters in the range 6.2-8.4nm had collapsed. Finally, a triple walled nanotube of 14nm internal diameter (the effective diameter is ~14.34m) was observed as collapsed too. All these 17 observations are in agreement with our theoretical predictions of eq. (2), supporting our conjecture of liquid-like nanotube bundles [1].

Nanotube number	Number N of walls	Diameter of the internal wall [nm]	Collapsed (Y/N) Exp. & Theo.
1	1	4.6	Y
2	1	4.7	Y
3	1	4.8	Y
4	1	5.2	Y
5	1	5.7	Y
6	2	4.2	N
7	2	4.6	N
8	2	4.7	N
9	2	6.2	Y
10	2	6.5	Y
11	2	6.8	Y
12	2	6.8	Y
13	2	7.9	Y
14	2	8.3	Y
15	2	8.3	Y
16	2	8.4	Y
17	3	14.0	Y

Table 1: Self-collapse of nanotubes in a bundle: our theory exactly fits the experimental observations [1].

Sliding strength

Assuming sliding failure, the energy balance during a longitudinal delamination (here "delamination" has the meaning of Mode II crack propagation at the interface between adjacent nanotubes) dz under the applied force F, is:

$$d\Phi - Fdu - 2\gamma(P_C + P_{vdW})dz = 0 \qquad (3)$$

where $d\Phi$ and du are the strain energy and elastic displacement variation due to the infinitesimal increment in the compliance caused by the delamination dz; P_{vdw} describes the still existing van der Waals attraction (e.g. attractive part of the Lennard-Jones potential) for vanishing nominal contact nanotube perimeter $P_C = 6a$ (the shear force between two graphite single layers becomes zero for nominally negative contact area); $6a$ is the contact length due to polygonization of nanotubes in the bundle, caused by their surface energy γ. Elasticity poses $\frac{d\Phi}{dz} = -\frac{F^2}{2ES}$, where S is the cross-sectional surface area of the nanotube, whereas according to Clapeyron's theorem $Fdu = 2d\Phi$. Thus, the following simple expression for the bundle strength ($\sigma_C = F_C/S$, effective stress and cross-sectional surface area are here considered; F_C is the force at fracture) is predicted:

$$\sigma_C^{(theo)} = 2\sqrt{E\gamma \frac{P}{S}} \qquad (4)$$

in which it appears the ratio between the effective perimeter ($P = P_C + P_{vdW}$) in contact and the cross-sectional surface area of the nanotubes.

Assuming a non-perfect alignment of the nanotubes in the bundle, described by a non-zero angle β, the longitudinal force carried by the nanotubes will be $F/\cos\beta$, thus the equivalent Young' modulus of the bundle will be $E\cos^2\beta$, as can be evinced by the corresponding modification of the energy balance during delamination; accordingly:

$$\sigma_C = 2\cos\beta\sqrt{E\gamma \frac{P}{S}} \qquad (5)$$

The maximal achievable strength is predicted for collapsed perfectly aligned (sufficiently overlapped) nanotubes, i.e. $\frac{P}{S} \approx \frac{1}{Nt}$, where t is the graphene thickness, $\beta = 0$:

$$\sigma_C^{(theo,N)} = 2\sqrt{\frac{E\gamma}{Nt}} \qquad (6)$$

Taking $E = 1\text{TPa}$ (Young's modulus of graphene), $\gamma = 0.2\,\text{N/m}$ (surface energy of graphene; however note that in reality γ could be also larger as a consequence of additional dissipative mechanisms, e.g. fracture and friction in addition to adhesion), the predicted maximum strength for single walled nanotubes (N=1) is:

$$\sigma_C^{(\max)} = \sigma_C^{(theo,1)} = 48.5\,\text{GPa} \tag{7}$$

whereas for double or triple walled nanotubes $\sigma_C^{(theo,2)} = 34.3\,\text{GPa}$ or $\sigma_C^{(theo,3)} = 28.0\,\text{GPa}$. Eq. (7) suggests the feasibility of 30MYuri strong tethers.

Self-buckling and sliding strength coupling

According to the previous analysis, the ratio between the bundle strength $\sigma_C^{(0)}$, in the presence of self-collapse, and $\sigma_C^{(O)}$, in the absence of self-collapse, is predicted to be:

$$\frac{\sigma_C^{(0)}}{\sigma_C^{(O)}} = \sqrt{\frac{2\pi R + P_{vdW}}{2\pi R\left(1 - \frac{1}{R}\sqrt{\frac{N^\alpha D}{2\gamma}}\right) + P_{vdW}}}, \text{ for } R \geq R_C^{(N)} = \sqrt{\frac{3N^\alpha D}{\gamma}} \tag{8}$$

The maximal strength increment induced by the self-collapse is thus:

$$\left.\frac{\sigma_C^{(0)}}{\sigma_C^{(O)}}\right|_{\max} = \sqrt{\frac{1}{1 - \frac{1}{\sqrt{6}}}} \approx 1.30 \tag{9}$$

Eq. (9) shows that the self-collapse could enhance the nanotube bundle strength up to ~30%. The reason is obviously the incremented surface area of the interfaces between the nanotubes.

Conclusions

The calculation in eq. (7) suggests a maximal achievable strength larger than 30MYuri, thus compatible with the Artsutanov's dream of the space elevator. Strong adhesion energy, high stiffness, low fiber dimension (thus aggregation must be avoided) and high alignment are all key factors for a practical realization of the single walled nanotube super strong bundle. Graphene bundles are expected to be even stronger [26].

References

[1] N. Pugno, The design of self-collapsed super-strong nanotube bundles. J. of the Mechanics and Physics of Solids, 2010, 58, 1397–1410.

[2] Z. Wu, Z. Chen, X. Du, J. M. Logan, J. Sippel, M. Nikolou, K. Kamaras, J. R. Reynolds, D. B. Tanner, A. F. Hebard, A. G. Rinzler, Transparent conductive carbon nanotube films, Science, 2004, 305, 1273–1273.

[3] M. Endo, H. Muramatsu, T. Hayashi, Y. A. Kim, M. Terrones, M. S. Dresselhaus, Nanotechnology: 'Buckypaper' from coaxial nanotubes, Nature, 2005, 433, 476–476.

[4] S. Wang, Z. Liang, B. Wang, C. Zhang, High-strength and multifunctional macroscopic fabric of single-walled carbon nanotubes, Advanced Materials 2007, 19, 1257–1261.

[5] M. Zhang, S. Fang, A. A. Zakhidov, S. B. Lee, A. E. Aliev, C. D. Williams, K. R. Atkinson, R. H. Baughman, Strong, transparent, multifunctional, carbon nanotube sheets, Science 2005, 309, 1215–1219.

[6] H. W. Zhu, C. L. Xu, D. H. Wu, B. Q. Wei, R. Vajtai, P. M. Ajayan, Direct synthesis of long single-walled carbon nanotube strands, Science 2002, 296, 884–886.

[7] K. Jiang, Q. Li, S. Fan, Nanotechnology: Spinning continuous carbon nanotube, Nature 2002, 419, 801–801.

[8] A. B. Dalton, S. Collins, E. Munoz, J. M. Razal, Von H. Ebron, J. P. Ferraris, J. N. Coleman, B. G. Kim, R. H. Baughman, Super-tough carbon-nanotube fibres, Nature 2003, 423, 703–703.

[9] L. M. Ericson, H. Fan, H. Peng, V. A. Davis, W. Zhou, J. Sulpizio, Y. Wang, R. Booker, J. Vavro, C. Guthy, A. N. G. Parra-Vasquez, M. J. Kim, S. Ramesh, R. K. Saini, C. Kittrell, G. Lavin, H. Schmidt, W. W. Adams, W. E. Billups, M. Pasquali, W.-F. Hwang, R. H. Hauge, J. E. Fischer, R. E. Smalley, Macroscopic, neat, single-walled carbon nanotube fibers, Science 2004, 305, 1447–1450.

[10] M. Zhang, K. R. Atkinson, R. H. Baughman, Multifunctional carbon nanotube yarns by downsizing an ancient technology, Science 2004, 306, 1358—1361.

[11] Y.-L. Li, I. A. Kinloch, A. H. Windle, Direct spinning of carbon nanotube fibers from chemical vapor deposition synthesis, Science 2004, 304, 276–278.

[12] K. Koziol, J. Vilatela, A. Moisala, M. Motta, P. Cunniff, M. Sennett, A. Windle, High-performance carbon nanotube fiber, Science 2007, 318, 1892–1895.

[13] K. S. Novoselov, A. K. Geim, S. V. Morozov, D. Jiang, Y. Zhang, S. V. Dubonos, I. V. Grigorieva, A. A. Firsov, Electric field effect in atomically thin carbon films, Science 2004, 306, 666–669.

[14] C. Berger, Z. Song, X. Li, X. Wu, N. Brown, C. Naud, D. Mayou, T. Li, J. Hass, A. N. Marchenkov, E. H. Conrad, P. N. First, W. A. de Heer, Electronic confinement and coherence in patterned epitaxial graphene, Science 2006, 312, 1191–1196.

[15] S. Stankovich, D. A. Dikin, G. H. B. Dommett, K. M. Kohlhaas, E. J. Zimney, E. A. Stach, R. D. Piner, S. T. Nguyen, R. S. Ruoff, Graphene-based composite materials, Nature 2006, 442, 282–285.

[16] D. A. Dikin, S. Stankovich, E. J. Zimney, R. D. Piner, G. H. B. Dommett, G. Evmenenko, S. T. Nguyen, R. S. Ruoff, Preparation and characterization of graphene oxide paper, Nature 2007, 448, 457–460.

[17] N. Pugno, R. Ruoff, Quantized Fracture Mechanics, Philosophical Magazine, 2004, 84/27, 2829-2845.

[18] N. Pugno, Dynamic Quantized Fracture Mechanics. Int. J. of Fracture, 2006, 140, 159-168.

[19] N. Pugno, New quantized failure criteria: application to nanotubes and nanowires. Int. J. of Fracture, 2006, 141, 311-323.

[20] N. Pugno, Young's modulus reduction of defective nanotubes. Applied Physics Letters 2007, 90, 043106-1/3.

[21] N. M. Pugno, On the strength of the nanotube-based space elevator cable: from nanomechanics to megamechanics. J. of Physics - Condensed Matter, 2006, 18, S1971-1990.

[22] N. M. Pugno. The role of defects in the design of the space elevator cable: from nanotube to megatube. Acta Materialia, 2007, 55, 5269-5279.

[23] N. M. Pugno, Space Elevator: out of order?. Nano Today, 2007, 2, 44-47.

[24] J. A. Elliott, J. K. Sandler, A. H. Windle, R. J. Young, M. S. Shaffer, Collapse of single-wall carbon nanotubes is diameter dependent. Physical Review Letters, 2004, 92, 095501.

[25] M. S. Motta, A. Moisala, I. A. Kinloch, A. H. Windle, High performance fibres from 'Dog Bone' carbon nanotubes, Advanced Materials, 2007, 19, 3721-3726.

[26] N. Pugno, Towards the Artsutanov's dream of the space elevator: the ultimate design of a 35GPa stronger tether thanks to graphene, Acta Astronautica, 2013, 82, 221-224.

SATELLITE PLACEMENT USING THE SPACE ELEVATOR

Stephen S. Cohen

CEGEP Vanier College, Professor, Physics Department, 821 Sainte-Croix, Montreal, QC H4L 3X9 Canada

Arun K. Misra

McGill University, Thomas Workman Professor and Chairman, Department of Mechanical Engineering, 817 Sherbrooke Street West, Montreal, QC H3A 2K6 Canada

Abstract: The advent of a space elevator would ease the process of satellite placement by offering a more efficient and elegant avenue to space than do rockets. This paper describes the basic features associated with satellite placement via the space elevator such as the natural orbits available to launched satellites and the impulses required to circularize these orbits. The effects of an oscillating ribbon at the time of launch on the orbits reached are also determined. The energy costs associated with climber transit are described as well.

I. Introduction

The space elevator will consist of a ribbon extending from the surface of the Earth to a counterweight beyond the geostationary altitude, which will lie in the equatorial plane. Once this ribbon is deployed, climbers may ascend the ribbon to various altitudes, and release satellites into orbit. The process, when accomplished in this manner rather than by conventional rockets, is expected to be of the order of one hundred times less expensive.[1] The discovery of carbon nanotubes,[2] which might be a suitable material for the space elevator ribbon, has increased the likelihood that a space elevator will be constructed in the foreseeable future. With the progression of this technology, studying other aspects of the space elevator, such as satellite placement, becomes appropriate.

While some studies have obtained basic analytical and numerical results for the motion of the space elevator ribbon due to various excitations,[3-5] many of the details concerning satellite placement via the space elevator have not been examined yet. Such studies are important in revealing certain benefits and constraints associated with satellite placement in this manner. This paper considers the basic features associated with releasing satellites into orbit: first from a static ribbon, and then from an oscillating one. The energy costs of ascending the space elevator ribbon with a climber are also computed.

II. Satellite Placement

Figure 1 shows the main components of the space elevator. It consists of a ribbon, counterweight and climber, all in the equatorial plane of the Earth. The ribbon is in tension due to the gravity and centrifugal gradients, which act in opposite directions. The climber, propelled by an electric motor, will ascend the ribbon, transporting payloads to various altitudes.

Once a climber reaches a desired launch altitude, d_0, the satellite it contains may be released. If no additional velocity impulse is added to the satellite at its time of launch, it will have used no fuel to arrive in orbit. These orbits will be called 'free Earth orbits'; the only input energy required to place a payload in a free Earth orbit is that which powers the climber to its launch altitude (discussed in section III).

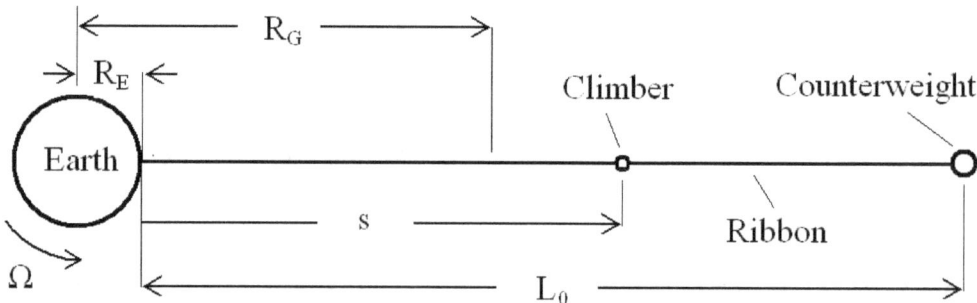

Figure 1. Schematic diagram of the space elevator

At the time of launch, let the climber (and the satellite it contains) have a radial distance r_0 from the center of the Earth, speed v_0, and flight path angle β_0. The values of v_0 and β_0 may be modified by an applied velocity impulse. The semi-major axis and eccentricity of the orbit that the satellite will fall into are given by (Ref. 6)

$$a = 1/(2/r_0 - v_0^2/\mu) \tag{1}$$

and

$$e = \sqrt{\left(\frac{r_0 v_0^2}{\mu} - 1\right)^2 \cos^2 \beta_0 + \sin^2 \beta_0} \tag{2}$$

respectively, where μ is the gravitational constant of the Earth.

A. Ideal Launch Scenario

Ideally, at the moment of satellite launch, the space elevator will be static (other than the nominal spin rate of the Earth), and in its nominal vertical configuration. If this is the case, and no additional impulse is applied to the satellite, then $r_0 = R_E + d_0$, $v_0 = \Omega(R_E + d_0)$ and $\beta_0 = 0$, where R_E and Ω are the radius and angular velocity of the Earth, respectively. The resulting semi-major axis and eccentricity pairings are plotted in Fig. 2 with all lengths non-dimensionalized with respect to R_E. It is observed that the geosynchronous altitude is the only point on the ribbon that yields a natural circular orbit ($e = 0$, shown by a bullet in Fig.2). If a mass is released from any other altitude in the range given, it will be in an elliptical orbit.

The minimum launch altitude considered is 23,500 km (3.685 Earth radii), because at any point below this, the perigee of the elliptical orbit of the satellite will be within 100 km from the surface of the Earth. The maximum launch altitude considered here is 42,000 km (6.585 Earth radii), because the semi-major axis of the resulting orbit for launches at altitudes beyond this point is very large.

Though not shown in Fig. 2, $d_0 = 7.345 R_E$, or about 46,850 km, is a critical launch altitude, which places payloads in a parabolic orbit. Therefore, for satellite placement in free Earth orbits, the launch altitude will be in the range given by $23{,}500 < d_0 < 46{,}850$ km. Also, any natural ($\Delta v = 0$) launch in the range given by $46{,}850$ km $< d_0 < L_0$, where L_0 is the total length of the ribbon, will send the payload into a hyperbolic orbit, as it will have a velocity greater than the escape velocity given by:

$$v_{esc} = \sqrt{2\mu/r_0} \qquad (3)$$

These trajectories are the starting point for planning interplanetary space missions using the space elevator, but are not discussed further, as this paper is limited to the placement of satellites into Earth orbits.

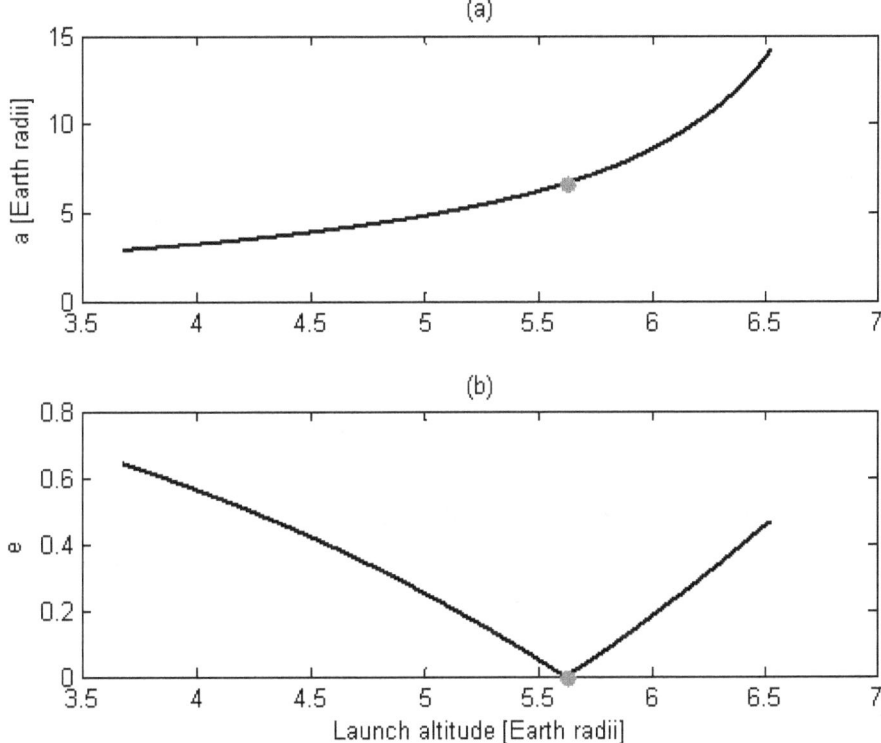

Figure 2. Orbit parameters vs. launch altitude: *(a) semi-major axis, (b) eccentricity*

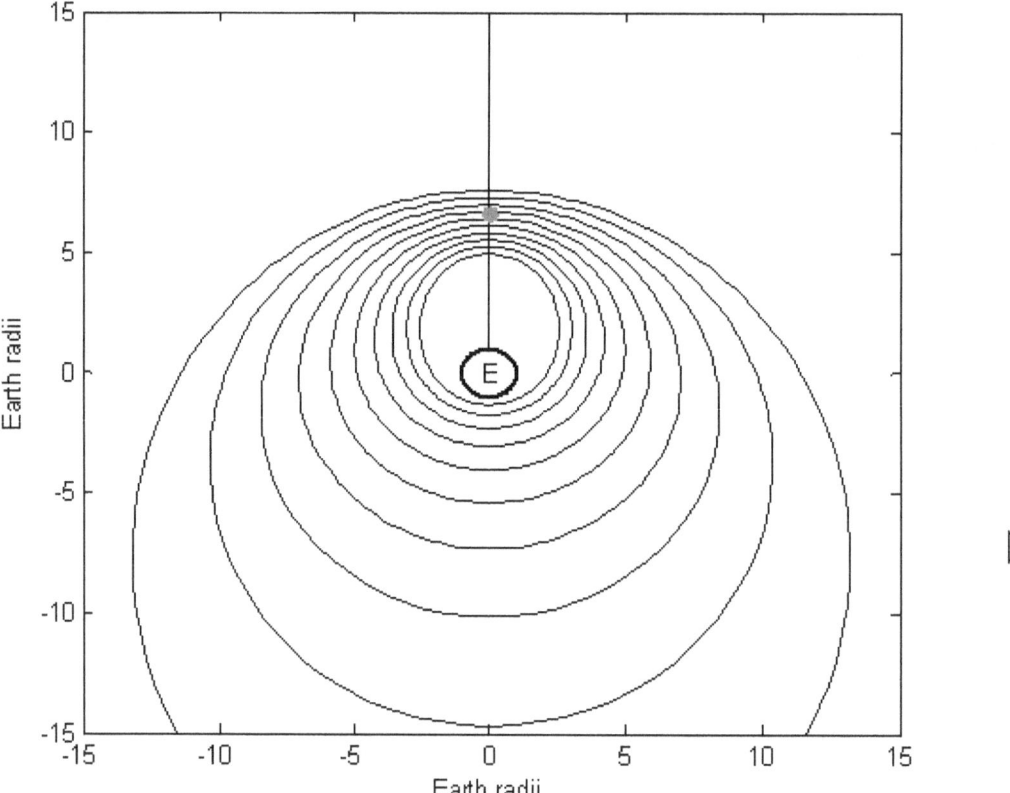

Figure 3. Free Earth orbits using the space elevator

A better picture of what has been called the free Earth orbits available to satellites using the space elevator is shown in Fig. 3. Since the flight path angle of the climber at the time of launch is zero, the point of launch can only be the apogee or the perigee of the orbit. Since the portion of ribbon below R_G is traveling slower than it would in a natural circular orbit, launches below this radius commence at the apogee of the orbit. Conversely, for launches above R_G, the initial radius corresponds to the perigee of the orbit.

While Fig. 3 shows the spectrum of free orbits available to Earth satellites, there are a wide range of reasonably low cost orbits that may be reached by transferring from the free orbits with a small impulse, Δv. For example, if a particular *circular* orbit of radius r_c having $r_c \neq R_G$ (where R_G is the geosynchronous radius) is desired, a particular Δv will be required. The most efficient elliptical to circular orbit transfer occurs at the perigee of an elliptical orbit if $r_c < R_G$, and at its apogee if $r_c > R_G$. These transfers are equivalent to the second impulse of a Hohmann transfer. For the case of $r_c < R_G$, the required impulse for circularization is given by

$$\Delta v = \sqrt{\frac{\mu}{a}} \left(\sqrt{\frac{1}{1-e}} - \sqrt{\frac{1+e}{1-e}} \right) \qquad (4)$$

where a and e are the semi-major axis and eccentricity of the original elliptical orbit, respectively. For the case of $r_c > R_G$, the required impulse is given by

$$\Delta v = \sqrt{\frac{\mu}{a}} \left(\sqrt{\frac{1}{1+e}} - \sqrt{\frac{1-e}{1+e}} \right) \quad (5)$$

It is noted that in *either* case, the impulse manoeuvre takes place at the side of the orbit opposite

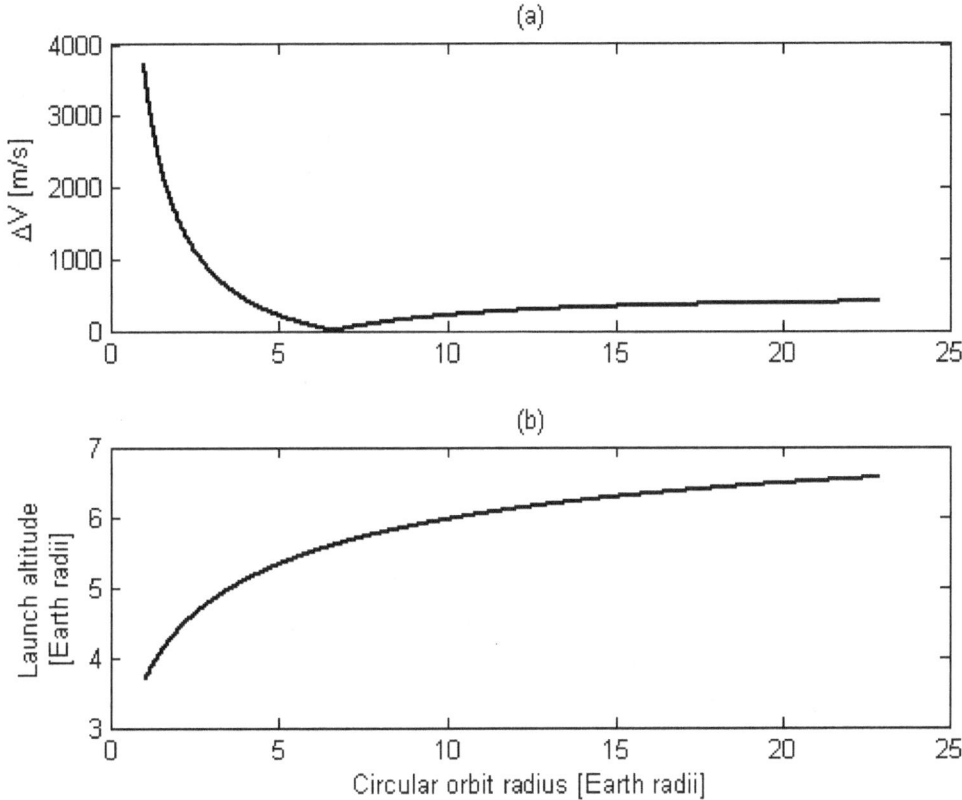

Figure 4. Minimum impulse required (a) and original launch altitude (b) to arrive in circular orbits of various radii

that which it was originally released from the climber.

In Figure 4, the minimum required impulse and the launch altitude of the original elliptical free Earth orbit are plotted against a wide range of desired circular orbits. It is noted that the required impulse for a geosynchronous orbit is zero, as it should be. The only circular orbits that require large impulses are those at low altitudes (altitude in the hundreds of kilometres). The benefits of using the space elevator to reach orbits in this range are limited; the required impulse using the space elevator as a platform actually exceeds 3 km/s. One potential way to reach circular low-Earth orbits via the space elevator economically is to release satellites from slightly below the 23,500 km altitude with the intent to aero-brake in the upper atmosphere of the Earth on its near pass. If such a manoeuvre is not feasible, the space elevator will probably not be used to transport satellites to circular orbits of such low altitudes, as rockets can accomplish this task rather inexpensively.

The benefits of launching satellites into circular Earth orbits with the space elevator instead of rockets become significant for larger orbits. Every circular orbit having a radius in the range given by 30,300 < r_c < 43,500 km may be arrived at with an impulse of less than 100 m/s.

B. Non-Ideal Launch Scenario (Oscillating Ribbon)

Now, the fact that the space elevator may not be static at the time of launch is taken into account. As shown in Ref. 5, a residual oscillatory rotation of the space elevator's ribbon, or, 'libration angle' α, measured from the local vertical at the surface of the Earth, may be introduced to a static space elevator by the Coriolis acceleration of a climber. In the case of climber transit, the amplitude of the residual libration, α_{res}, will likely be in the range of milliradians. It is unlikely that any other excitation will cause oscillations of an order higher than this. When the ribbon undergoes steady state vibration, its maximum libration rate takes place as it passes the vertical position (α = 0). This is, however, an ideal moment to release a payload because it coincides with the instant when its flight path angle is zero. The libration rate at this moment is given by $\Omega W \alpha_{res}$, where W is the non-dimensional natural frequency of the ribbon.[7] The perturbation velocity caused by this libration rate, which points in the transverse direction and is, of course, equal to the required impulse to counter it, is given by

$$\Delta v = \Omega W \alpha_{res} d_0 \qquad (6)$$

The product ΩW will have a value of about 1.5×10^{-5} for expected ribbon material and design parameters. Also, as already mentioned, for satellite placement, 23,500 < d_0 < 46,850 km. So, if a satellite is launched from a ribbon that is oscillating with an amplitude of the order of milliradians at the moment when it is vertical, applying the correct reasonably small impulse (of the order of metres per second) in the opposite direction will eliminate the effect of oscillation.

Finally, as shown in Figure 5, if the libration on the order of milliradians is simply ignored, and a satellite is launched from the geosynchronous altitude at the moment when the ribbon is vertical, the semi-major axis of the orbit of the satellite will change by tens of kilometers, and its eccentricity by the order of 10^{-3}. If the ribbon is not vertical at the time of launch, then in addition to the perturbations experienced by the two aforementioned orbital parameters, the argument of the perigee of the orbit will be slightly modified, because $\beta_0 \neq 0$.

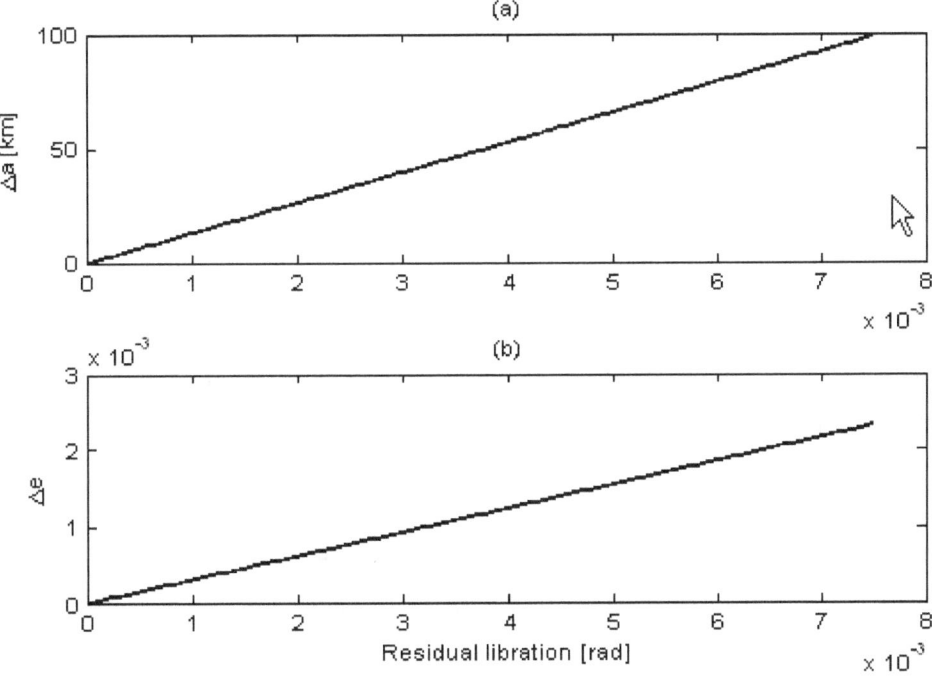

Figure 5. Deviation in (a) semi-major axis and (b) eccentricity of the geosynchronous circular orbit due to oscillating ribbon for launch at moment when the ribbon is vertical

III. Operational Costs for Climber Transit

From a business standpoint, the primary motivation for the implementation of a space elevator is that it would drastically reduce the cost of satellite placement and space missions. Its construction costs, which are heavily dependent upon the cost of carbon nanotube synthesis, have been approximated in the tens of billions of dollars (USD).[1] While this may seem expensive, with frequent use, the savings incurred during the space elevator's operation could be enough to recover these setup costs rather quickly. The free Earth orbits, which have been described above, are orbits that require no energy input other than that required to transport the climber and payload to its launch altitude. This section aims to determine these energy costs.

The case where a payload is carried inside a climber from the surface of the Earth to some altitude, h, below the geosynchronous orbit, is examined. The climb beyond this altitude is propelled by the spin of the Earth, the effect of which is greater than that of the gravitational force beyond the geosynchronous altitude. If the climb is done at a constant speed, and the ribbon remains in its nominal position, the required thrust by the climber, F_T, is given by

$$F_T = m_e \left[\mu / (s+R_E)^2 - \Omega^2 (s+R_E) \right] \tag{7}$$

where m_e and s are the mass (including payload) and altitude of the climber, respectively. The work done by the climber, W_T, is found by integrating the product of climber thrust and differential climber position:

$$W_T = \int F_T ds \tag{8}$$

The work per unit mass done by the motor to move the climber from the surface of the Earth to h is then given by

$$\frac{W_T}{m_e} = \frac{\mu}{R_E}\left(\frac{h}{R_E+h}\right) - \frac{\Omega^2}{2}\left[(R_E+h)^2 - R_E^2\right] \tag{9}$$

Therefore, the work per unit mass done by the motor to move the climber from the surface of the Earth to the geosynchronous altitude is about 48.5 MJ/kg. An apparent anomaly occurs when examining the change in the energy per unit mass of the climber before and after transit, $E_{1\to 2}$, given by

$$\frac{E_{1\to 2}}{m_e} = \left[\frac{\Omega^2}{2}(R_E+h)^2 - \frac{\mu}{(R_E+h)}\right] - \left[\frac{\Omega^2}{2}R_E^2 - \frac{\mu}{R_E}\right] \tag{10}$$

or

$$\frac{E_{1\to 2}}{m_e} = \frac{\mu}{R}\left(\frac{h}{R_E+h}\right) + \frac{\Omega^2}{2}\left[(R_E+h)^2 - R_E^2\right] \tag{11}$$

The increase in the total energy per unit of mass of the climber in moving from the surface of the Earth to the geosynchronous altitude is about 57.7 MJ/kg. This result begs the questions, "How does the climber gain more energy than the work that the motor puts into it (48.5 MJ/kg)?" and "Where does the free energy come from?" The *free* energy comes from the diurnal rotational energy of the Earth. For a transit to the altitude h, the energy per unit mass extracted from the Earth is given by $\Omega^2[(R_E+h)^2 - R_E^2]$; for the particular case of transit to geosynchronous altitude, $h = R_G - R_E$, so the energy per unit mass extracted from the Earth's spin is $\Omega^2(R_G^2 - R_E^2)$ or 9.2 MJ/kg. The slowing of the spin of the Earth due to this energy extraction is negligible for any reasonable amount of mass that is transported. The situation is analogous to that of gravity assist. What this result implies is that even if a rocket were made to function with the same overall efficiency as that of the space elevator, it would still be less efficient, as it could not extract this free energy. It is however somewhat of a moot point, as these savings are small compared to those incurred due to the actual difference in the overall efficiency between the two methods of space travel.

The fuel cost for a typical shuttle launch to the geosynchronous orbit is around $210 per kilogram of payload.[1] The required energy cost per kilogram of payload in order to accomplish this same feat with the space elevator, C_{SE}, is given by

$$C_{SE} = 48.5\frac{C_e}{\eta \bar{M}_p} \tag{12}$$

C_e is the cost of electricity per MJ, and η is the overall efficiency of the energy conversion process, which includes that of the transmission from Earth via laser and that of the climber's motor. $\bar{M}_p$ is the ratio of the payload mass to that of the entire mass being lifted, m_e. If reasonable values are assumed for these parameters (C_e = 0.012 $/MJ, η = 0.3 and $\bar{M}_p$ = 0.65), the cost for transport to the geosynchronous orbit using the space elevator is around $3 per kilogram of payload.

It is clear that the space elevator could bring space travel costs down by two orders of magnitude. Also of interest is the fact that climbers ascending the ribbon at constant speed beyond the geosynchronous altitude would be required to brake, and could actually generate energy in doing so. A more comprehensive cost analysis of the space elevator is provided by Ref. 1.

IV. Conclusion

The basic features associated with satellite placement using the space elevator have been examined.

The spectrum of free Earth orbits available to satellites released from the climber is quite broad. Impulse manoeuvres will be required to reach any orbit not falling within this spectrum. With the exception of low-Earth orbits, such fuel costs will be *very* small compared to what is normally required for a rocket to attain such orbits without the aid of the space elevator. For example, although there is only one free circular Earth orbit (at the geosynchronous altitude), there is a wide range of low-cost (<100 m/s impulse) circular high-altitude orbits.

The fact that the ribbon will be in motion when the satellites are launched is not likely to play a major role in such launches. Since the steady state motion of the ribbon due to climber transit will be rotational oscillations of the order of milliradians, the effect of such dynamics may be countered by impulses of the order of metres per second at the time of payload release. Even if this ribbon motion were not accounted for at the time of release, its effect on the orbit of the satellite would be small: the semi-major axis would change by only tens of kilometres, and the eccentricity by the order of 10^{-3}.

Finally, the total energy cost incurred transporting mass from the surface of the Earth to the geosynchronous altitude is 48.5 MJ/kg. Thus, the implementation of a space elevator would decrease the cost of satellite placement, perhaps by two orders of magnitude, because unlike rockets, the climbers would carry no fuel and be powered by an electric motor.

References

[1] Edwards, B. C. and Westling, E. A., *The Space Elevator: A Revolutionary Earth-to-Space Transportation System*, ISBN 0972604502, published by the authors, 2003.

[2] Iijima, S., "Helical Microtubules of Graphitic Carbon," *Nature*, Vol. 354, 1991, pp. 56-58.

[3] Patamia, S. E. and Jorgenson, A. M., "Analytical Model of Large Scale Transverse Dynamics of Proposed Space Elevator," *56th International Astronautical Congress of the International Astronautics Federation*, Fukuoka, Japan, 2005, Paper No. IAC-05-D4.2.06.

[4] Lang, D. D., "Space Elevator Dynamic Response to In-Transit Climbers," *1st International Conference on Science, Engineering, and Habitation in Space*, Albuquerque, NM, 2006, Paper No. 10152148LANG.

[5] Cohen, S. S. and Misra, A. K., "The Effect of Climber Transit on the Space Elevator Dynamics," *Acta Astronautica*, Vol. 64, 2009, pp. 538-553.

[6] Kaplan, M. H., *Modern Spacecraft Dynamics & Control*, John Wiley & Sons, Inc, U.S.A., 1976.

[7] Cohen, S. S., "Dynamics of a Space Elevator," Masters Thesis, Dept. of Mechanical Engineering, McGill Univ., Montreal, Quebec, 2006.

SPACE ELEVATOR DEPLOYMENT

James G. Dempsey
Independent researcher, 85 Cove Lane, Oshkosh, WI 54902 USA
jim.00.dempsey@gmail.com

Abstract: A Space Elevator is an elastic structure, with a very low spring constant, which until anchored, does not obey the same orbital mechanics rules developed for rigid bodies. This article will address the issues related to deployment and offer a suggestion for resolving these issues.

V. Introduction

This document is derived from the U.S. Patent 7,971,830, July 5, 2011 "SYSTEM AND METHOD FOR SPACE ELEVATOR DEPLOYMENT", inventor: James G. Dempsey.

The author is an independent researcher who has had an interest in space elevator since 1968. In 2003 the author has begun extensive computer simulations relating to space elevators.

VI. Orbiting Very Large Elastic Structures

The body of knowledge relating to the field of orbital mechanics, has been developed for use with rigid bodies. While some of the bodies may include flexible components, such as solar panels, the orbital paths are computed as if theses bodies were rigid. Relatively small elastic structures, on the order of a few 100 meters to a few kilometers, will have an aggregate orbital characteristic of that of a rigid body. A space elevator is an exception to this quasi-rigid body.

During deployment and prior to anchorage a space elevator will become an orbiting elastic structure on the order of 100,000 kilometers in length. This type of structure has not been placed in orbit other than through simulation studies. To date, there is no real world experience in orbiting such a structure. Simulation studies performed by this author demonstrate that a new skill set needs to be developed for deploying very long orbiting elastic structures.

A few simple thought experiments with familiar scenarios will provide you with some insight as to the problems involved.

Experiment 1: A fishing pole reel wound with steel wire affixed to a purchase point 1 meter above a steel plate and with a small magnet attached to the loose end of the steel wire. The object of the experiment is to slowly lower the magnet to the steel plate while measuring the tension force in the wire. You will notice that as you lower the magnet you can maintain a relatively controlled rate of descent. The tension force in the wire remains relatively constant until the magnet gets within a few centimeters of the steel plate. At this point the change in tension force per distance of magnet to steel plate is noted as approximately inversely proportional to the square of the distance of the magnet to the plate. Additionally, due to the relatively short length of steel wire, and the relatively high spring constant, the amount of strain (and resultant elongation) is relatively small. And thusly, with this relatively small strain/elongation, the rate of descent of the magnet can be controlled with a reasonable amount of accuracy by varying the deployment rate at the reel end.

Experiment 2: Same setup as experiment 1 excepting that the reel is loosely wound (but snug) with a thin elastic band. Now you run the deployment test you notice that as you unwind the elastic band that the spring constant diminishes inversely with the length of the unwound elastic band. In order to maintain a fixed descent rate of the magnet, you have to decrease the deployment rate of the elastic band as you deploy. Later in deployment, when the magnet approaches the steel plate, you find the increase in magnetic force as the magnet approaches the steel plate exceeds that of the increase of spring force of the elastic band. Meaning that at some point during deployment you would have to start reeling in elastic band to maintain equilibrium while attempting to maintain the constant descent rate of the magnet. You will likely conclude that you cannot manually control the descent rate nor place the magnet at a relatively close distance to the plate. I will also venture to guess, that even with computer control, with very responsive motor control on the reel, that you will also find it next to impossible to control the descent rate through the final few centimeters of descent.

Now let's extrapolate the experience learned from the second thought experiment, and thus try to anticipate the problems involved with deployment. Let's assume that the choice of deployment is to place the appropriate amount of tether, on two spools, into GEO orbit above the desired anchorage point. The intention is to simultaneously deploy the outbound spool and inbound spool at controlled rates such that the GEO position of the tether is maintained. This can only be accomplished by either a) precisely controlling the tension differential at the GEO position through exceptional control of deploying ends, or b) compensating for tension differential with the use of thrusters at the GEO position.

Unlike experiment 2, your purchase point is not fixed. Any imbalance in the deployment tensions at the GEO position will require an external force applied to maintain position. This force would presumably come from thrusters attached to the tether at GEO and/or at the deploying spools. This author anticipates that it will be advantageous to keep the mass of the deploying ends at minimum. Meaning any external force (thrust) will be applied at the GEO position of the deploying tether.

An additional factor when trying to control the tensions at the spool ends by way of deployment rate and/or thrusters is that any effect from this control cannot be transmitted through the tether at a speed faster than the speed of sound for the material. It is unknown as to the eventual speed of sound for a yet to be built space elevator tether but for metals it is on the order of 5,000 meters per second. Depending on which end applies the control, the GEO position could not experience the effect for a delay of upwards of 5-15 seconds.

Positions along the portion of the deployed tether, together with the un-deployed tether and spools experience two primary forces: a) centripetal force that is proportional to the radius from center of earth, and b) gravitational force that is inversely proportional to the square radius from center of earth. The net force, at any point along the deployed portion of the tether and at each spool with un-deployed tether is the sum of these two forces. The deploying ends will also experience the Coriolis Effect causing the outward bound end to tilt westward during deployment, and the downward bound end to tilt eastward during deployment. The Gravitational and centripetal forces will eventually supply sufficient torque to eventually right the tether.

The object of the deployment is to maintain the station position of the GEO station of the tether without requiring an inordinately large amount of fuel and to keep within a maximum thrust budget.

Complete control of the differential tensions at the GEO station, i.e. targeting a net zero force, together with reducing the time of deployment is paramount in reducing the fuel budget. Experiment

2, should tell you that the outbound end with diminishing gravitational force and increasing centripetal force together with diminishing spring constant will result in a deploying end in which you can relatively control the tension at the GEO station. This is principally due to the sum of the forces approaching a linear distribution. Control of the outbound can be attained up until tether breakage which would not be a design feature of the deploy tether.

The inbound end of the deploying tether is a different story. As the inbound end approaches earth, the increase in gravitational force behaves very much like the increase in the magnetic field force of Experiment 2, and this together with the relatively low spring constant of the deployed tether representing that of the elastic band in Experiment 2, would result in, as you approach earth, a lack of control on the earth end of the deploying tether.

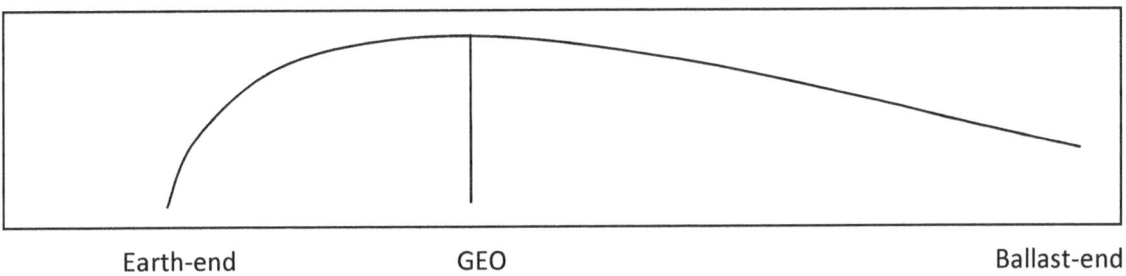

Earth-end   GEO   Ballast-end

Slope = Gravitational Acceleration (Ag) + Centripetal Acceleration (Ac)

Figure 1

The above curve, represents in the slope, the net effect of the combined acceleration experienced by a space elevator tether as contributed by the gravitational acceleration and the centripetal acceleration. The horizontal axis is the altitude (Earth-end 0, Ballast-end ~100,000 km). The vertical axis is a potential surface diagram with the slope indicating the direction of the net acceleration. Positive slope to the left (towards Earth), negative slope to the rights (towards Ballast-end).

Not depicted in the graph are Coriolis force, solar winds, light pressure, tides and other factors.

At Earth-end the slope is positive and quite steep indicating a preponderance of gravitational acceleration. At GEO, the slope is zero, indicating the gravitational acceleration equals the centripetal acceleration. At Ballast-end the slope is negative illustrating the centripetal acceleration exceeds that of the gravitational acceleration.

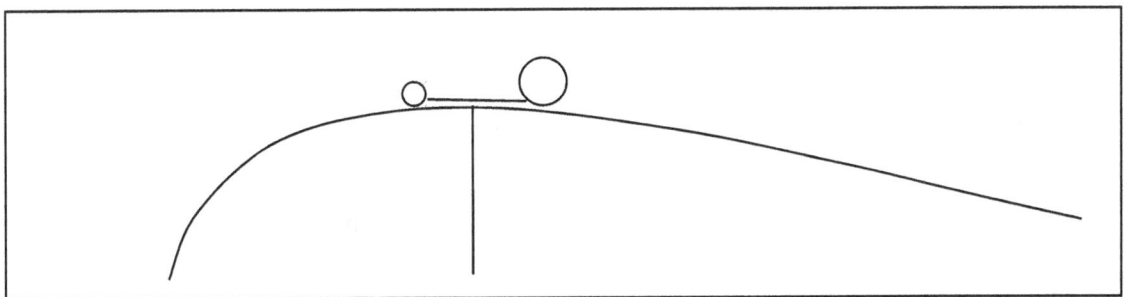

Tether deployment from two spools at GEO, early in the deployment

Figure 2

During the early stage of deployment, using dual spools from GEO, there are relatively small issues to resolve. The slope of the Ag + Ac curve is relatively complementary for each spool (with un-deployed tether mass) and along the tether deployed between the spools. The net acceleration experienced by the spools and tether is complementary. Additionally, the relatively short segment of deployed tether has a relatively high spring constant. This means that control forces at either spool, both spools, or at a central point along the tether such as the GEO position, that the control force is immediately transmitted to the other components of the deploying tether system.

During the early phase of deployment, it is feasible to compute and control the deployment rates such that equilibrium is maintained in the deploying space elevator. If assistance is required to compensate for uncertainties, the application of thrusters at each spool or either spool and/or at GEO would provide effective control over and instabilities that may be induced into the system by external forces.

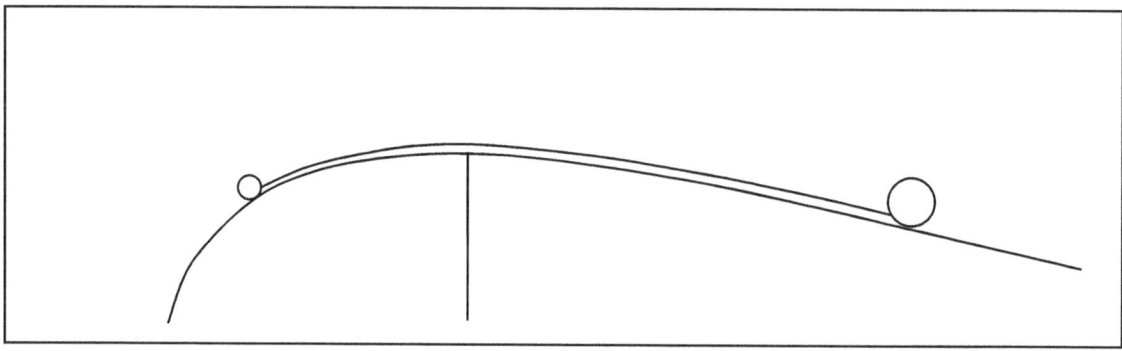

Tether deployment from two spools at GEO, midpoint in the deployment

Figure 3

As deployment progresses, maintaining system equilibrium, becomes increasingly difficult. The principal factors affecting the equilibrium solution are:

- Net accelerations experienced by deployed tether
- Tension along the deployed tether
- Changing mass of each un-deployed tether spool and deployment mechanism
- Net acceleration (Ag + Ac) of each un-deployed tether spool and deployment mechanism
- Velocity of each spool along the Ag + Ac slope
- Acceleration (change in deployment velocity) of each spool
- Acceleration (change in radial velocity) of each spool
- Diminishing spring constant of deployed tether
- Movement of localized tether mass due to low spring constant and attempts at control

Note, the above factors have not listed the Coriolis force effects, the effect of movement of tether mass as a result of deployment rate change, tidal effects, solar wind, light pressure and other factors complicating the equilibrium solution.

Due to the number of variables affecting the equilibrium solution, any small change in any one variable quickly multiplies the changes required in the other variables necessary to maintain equilibrium. The two variables contributing the most severe consequences to the equilibrium solution is the diminishing spring constant of the deployed tether combined with the rate change of slope of curve (related to velocity of each spool).

In attempting to maintain precise control in the radial positions of each spool, and considering the disparate slopes of the $A_g + A_c$ curve, the deployment velocities must be precisely varied and varied at different rates. Changing the deployment velocity alone will affect the tether tension, first at the spool end of the tether, then propagating as a tension wave along the length of the tether. This tension wave moves tether mass as the wave propagates along the tether, and eventually the other spool mass is moved in response. Computing the mass flow effect adds an additional variable to the equilibrium equation and thus complicates the solution.

To compensate for this, one might consider adding thrusters and fuel tanks to the deployment mechanisms. Unfortunately this adds mass to the deploying ends of the tether, which happens to be located at the position of the highest rate change in slope. The additional mass may require an increase in initial tether load bearing capacity. This increases the initial mass (and cost) required to be lifted to orbit. The additional mass exacerbates the equilibrium solution thus requiring more fuel and at some point higher rated thrusters.

An attempt to compensate for shortcomings of the first corrective procedure would be to consider placing additional thrusters and fuel tanks at GEO. There are problems with this tactic:

- The very low spring constant, late in deployment, requires a large displacement in tether material to affect a reasonable force on the spools.
- An attempt to compensate for one spool position/velocity affects the other spool
- There is a time delay between applying the control force and having the spool experience the tension force.
- The control attempts on the un-damped spring nature of the tether tends to come self-reinforcing (Lang study).
- Thrusters (e.g. Hydrazine) are not infinitely variable and tend to be pulsed on and off introducing small velocity changes into a much larger spacecraft or satellite. Other than for the mass of the thruster and fuel tank there is virtually no mass of the tether at the point of where thruster applies its force.

Pulsing the thruster will tend to transmit tension pulses along the tether in both directions but of opposite magnitude. Attempted control of one spool by means of thruster at GEO will affect the other spool as well. The spring action of the tether will tend to absorb the thrust pulse and then produce a restorative force once the thrust is turned off. Only by displacing the tether, then holding the displacement for a significant duration of time, i.e. approximate time for tension wave to traverse to the spool, would you be able to transmit a significant control force to the spools. This requires longer thruster burn time (more fuel), and/or increase in number of burn pulses. Thrusters have limited life in number of on/off cycles which have to be factored into the solution equation.

The lack of control of the earth end of the deploying tether would result in requiring additional compensation thrust at the GEO station. Failure to accurately control the thrust requirements to maintain control over the deploying spools positions and rates would result in either an earth crash of the space elevator or an uncontrolled launch of the space elevator into a higher orbit. Some simulation studies have shown launching to an orbit with an apogee beyond that of the moon.

The following two simulation runs illustrate uncontrolled deployment scenarios.

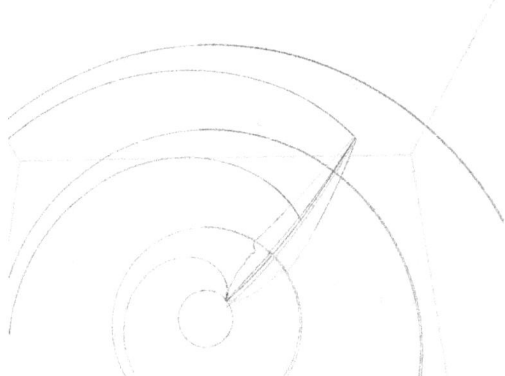

Uncontrolled crash resulting from deployment ~1 ft below GEO equilibrium point.

Figure 4

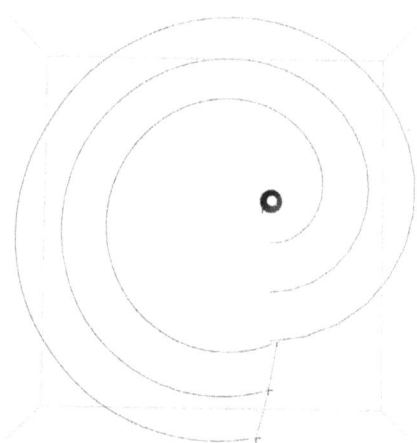

Uncontrolled flight into non-elliptical orbit resulting from deployment ~1 ft above GEO equilibrium.

Figure 5

In the above two simulation runs, the same computer controlled deployment algorithm was used to attempt to maintain equilibrium of deployment tensions as experienced at the GEO station (middle trace line in charts. Top chart does not show earth but does show anchor station trace line (circle). The bottom chart shows earth (North Pole).

An additional characteristic to learn of orbiting elastic structures is clearly illustrated in the second simulation run. This illustrates that the orbit of large elastic structures is not that of a rigid body. This is to say these orbits are not elliptical.

Author's note:

While it is unfounded to state it is impossible to deploy an initial space elevator of sufficient capacity to support a climber, it is evident to some degree that initial deployment of such a tether will require considerable care, may require excessive fuel, and may have a high probability of failure. As the fuel budget increases, the cost of lifting the fuel into orbit increases, and the initial load carrying capacity of the tether may increase, thus requiring additional costs of lifting additional tether mass, and in turn requiring additional control thruster fuel. The combination of interrelated factors results in a self-reinforcing cost spiral. This cost spiral, combined with the high probability of failure during the learning process of deploying tethers, results in an unacceptable risk. The conclusion is: a technique needs to be developed that minimizes the cost and risk factors to the point of acceptable risk.

VII. Controlling the Deployment

During the early years of rocketry, in the attempts to place satellites into orbit, we went through a learning process and a period of trial and error. Many failures were expected on the way to success. While we now use computer simulations to reduce the number of, and in some cases eliminate failures, failures must be expected and factored into the cost and risks of the initial use of any developing technology.

This author is exploring techniques to reduce the costs and associated risks of initial deployment to the point where a single failure, though not desirable, is not a "show stopper". The cost of a failure, or even a few failures, during the learning process, are to be reduced to the point where it would be acceptable considering the potential net return on a fully functioning space elevator.

To accomplish this we reduce the mass of tether, equipment, and fuel lifted to deployment level orbit. These reductions are made to the point where the initially deployed tether is self-supporting and has a minimal load bearing capacity. The only requirement of the initial load capacity is that it is useful for subsequent tether deployment from GEO in the build-up phase of deployment. This author is exploring techniques where the load carrying capacity of the tether, while insufficient to support a climber carrying additional tether, has the initial capacity sufficient to support additional tether deployment initiated from GEO. This permits the prior deployments to offer a purchase point and thus reduce or eliminate thruster fuel requirements for each subsequent build-up of the tether.

The first question to answer is: What is the minimal load carrying capacity to explore?

This author's thought process is to look at what is a reasonable low-to-mid-range control thruster capacity. Then size the tether load carrying capacity such that the initial tether has a similar load carrying capacity of that of the thruster. Once deployed, the initial tether's load carrying capacity can be used in lieu of, or to augment, control thrusters in the deployment of the secondary tether. It should be noted that the initial tether can supply a force to the secondary tether; equivalent to the thruster used for the initial deployment, but without requiring fuel, and can supply the force for an infinite period of time.

The thruster of choice, under examination by this author, is a hydrazine thruster, such as the EADS Astrium model CHT 20 with thrust range in the 7.9 to 24.6 N (1.776 to 5.53 lbf).

This thruster is on the order of 1/100th to 1/300th the capacity of the thrusters used in the Lang study.

Although the simulation studies of various configurations are in progress, the fuel minimizing techniques are being learned, and the proper initial tether profile is being determined, all indications are under 1000 kg of fuel will be required for initial deployment of a tether with a load carrying capacity in the range of 25 N.

As stated, the simulation studies are a work in progress. Wind effects studies on the initial tether have yet to be performed. It may be determined that a higher initial load carrying capacity is required to counteract wind forces. Additionally, the conclusion of the initial simulations may indicate a higher load carrying capacity can be attained within a reasonable fuel budget. Attaining a higher load carrying capacity of the initial tether would enable an increase in the load carrying capacity of the secondary tether and so on, and thus reduce the time to build up the tether to the point of supporting the initial climber.

Two of the deployment techniques under study by this investigator, and outlined within this report are:

- Dual spool from GEO – spool down/spool up deployment (bolo)
- Dual spool from GEO – loop down/loop up deployment (shoestring)

As compared with the traditional envisioning of deployment we have:

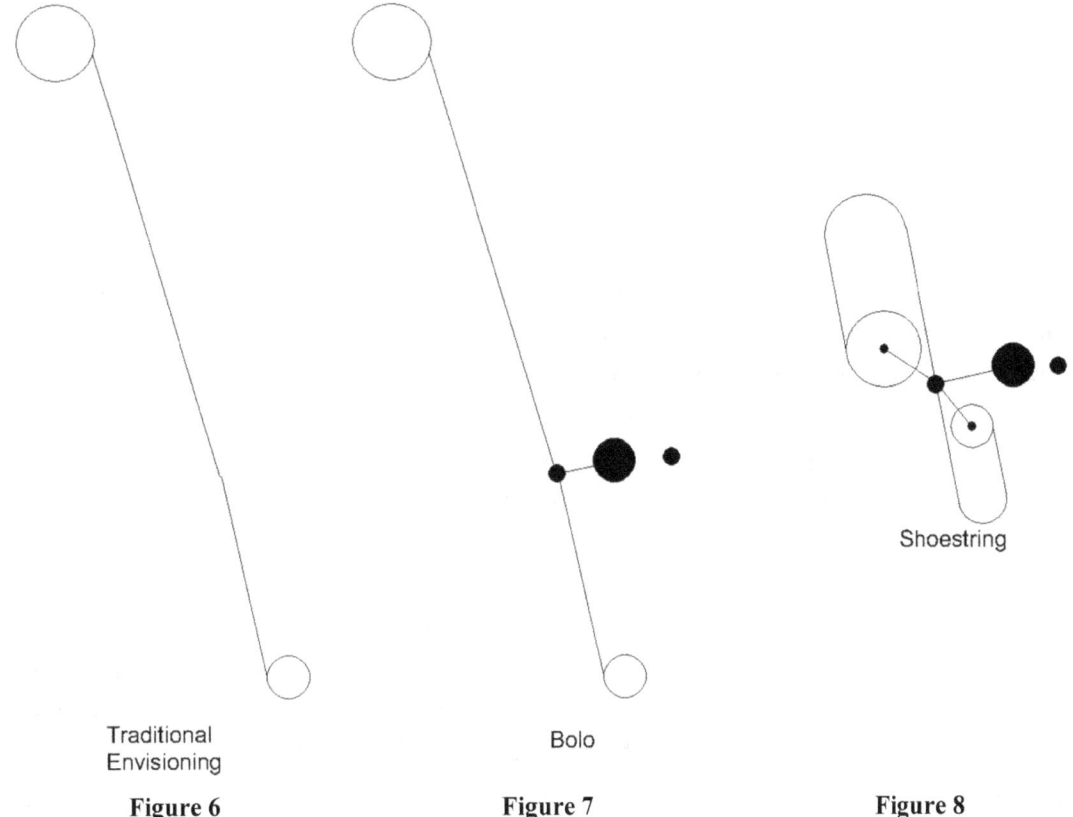

| Traditional Envisioning | Bolo | Shoestring |

Figure 6   **Figure 7**   **Figure 8**

Being an entrepreneur with an interest in space development, novel techniques discovered and developed in the process by this entrepreneur will seek Patent protection. The shoelace technique, and certain related aspects of the bolo technique are subject to a U.S. Patent 7,971,830, July 5, 2011.

The traditional envisioning of deployment (left in Figure 6) consists of deployment devices, primarily spools of tether with breaking systems to control the spool rate. Typically these systems are envisioned with a thrust control system and fuel placed with the top spool. Deployment begins at or below GEO and the thrust control system attempts to maintain equilibrium in the system as well as desired orbital position.

The two deployment techniques being presented in this paper are:

- Bolo technique (middle in Figure 7)
- Shoestring (right in Figure 8)

The principal differences between the Bolo and the Traditional Envisioning of deployment are:

- Addition of GEO ballast mass
- Addition of beacon satellite
- Movement of thrust control system to GEO ballast mass

In the Shoestring technique the spools are held captive to the GEO ballast mass and tether loops are deployed as opposed to the spools. The Shoestring techniques preserves as much of the tether mass near GEO until the (near) de-spooling of the tethers. The bottom loop contains tether of approximate length of the distance of Earth to GEO less strain, plus initial contortion (lower part of tether not initially strait upon anchorage). Upon de-spooling, the lower tether end, which is not attached to the spool, is left to "fall" into place.

Studies may indicate that the top end may optionally require thruster control. In which case, the top tether loop, which contains tether of a length of GEO to ultimate upper ballast altitude, is affixed to the spool and the spool is contained in a small device which also contains thrusters and fuel tank. As the top tether nears complete deployment the top tether spool (together with thrusters and fuel tank) is released from the GEO ballast object and left to "fall" upwards (outwards) while the breaking device on the top spool gently applies more breaking force to slow the de-spool rate such that the de-spool rate reaches zero when tether is depleted (or a short time before). Then the spool, with thrusters and fuel tank, remain attached to the tether as it falls outwards to the ballast altitude.

It is important to understand, the deployment rates and timing is such that the free "fall" of the ends of the tethers places the ends into the desired positions. The Earth end at or near anchor point, the upper ballast end at or near zenith above Earth end anchor point. The upper end will need thruster applied damping, but the Earth end, being anchored will not. Undulation in the tether during stabilization can be damped out using remaining fuel at the GEO ballast location on the tether and/or by manipulating the position of the anchor point.

As deployment progresses with the Shoestring technique the shape develops into that of a tied shoestring.

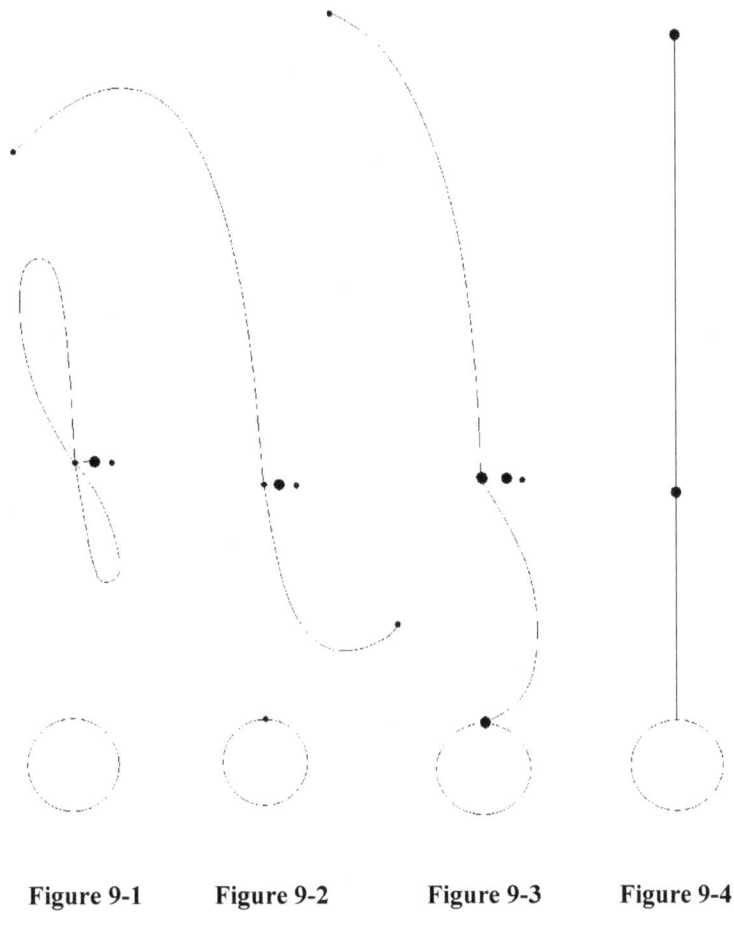

Figure 9-1   Figure 9-2   Figure 9-3   Figure 9-4

Shoestring deployment sequence

The sequence of Shoestring deployment from left to right

Figure 9-1   Partially deployed tether connected to GEO ballast object following beacon satellite
Figure 9-2   Post de-spooling of tethers with tether ends in flight
Figure 9-3   At time of Earth anchoring, prior to stabilization
Figure 9-4   Fully stabilized initial tether

Both the Bolo and Shoestring techniques have similar problems to solve and have taken the same solutions for the common problems.

The major problem shared by all initial deployment solutions is how to maintain control over the orbital altitude of the deploying orbiting structure as it nears completion. As noted by the previous studies presented at prior ISDC conferences, instabilities arise as the tether approaches the midpoint of deployment. And, these instabilities get worse as deployment progresses towards completion.

The two principal factors to the solution of the deployment problem are:

1. Keep the deploying mass down to a minimum (as compared to the total mass of the system) and thus minimizing the thrust requirement for corrective maneuvers. And 2) reduce the time to deployment.

To keep the deploying mass down, that is the mass migrating away from the GEO position, the preferred thruster and fuel tank position is placement at GEO. It is also significant that additional stability can be attained by having the ratio of the mass at GEO as compared with the mass of the tether and deployment devices and equipment away from GEO, be as large of ratio as possible. i.e. keep the majority of the mass at GEO during the periods of instability.

Under the traditional dual spool deployment (e.g. Lang and Dempsey studies and a deployment method proposed by Edwards) the mass is principally held in the spools and migrates away from GEO during deployment. To compensate for mass migration, it would be of benefit to place additional mass at the GEO initial deployment position, and attach this mass to the GEO position of the deploying tether. Then by controlling the orbit of the attached GEO ballast you control the general orbital characteristics of the total structure. In the Bolo and Shoestring technique the extra GEO ballast mass is depicted by the large dot connected the tether at the deployed tether GEO position represented by a smaller dot on the tether.

It should be noted that the momentum of this additional mass is important. This being, that should you simply use thrusters attached to the tether at the GEO position, and attempt control from there, then as the deployed tether length increases the spring constant decreases and additionally the length of time increases for a tension wave to propagate from the GEO position to a spool (or both spools). This increase in time for tension wave propagation, subject to lack of mass at GEO, would increase the burn time (or numbers of burn bursts) of the thrusters, and thus overburden the thrust control. Adding mass at GEO, and under control of the thrusters, reduces the number of burn bursts. The end effect is GEO ballast serves as an inertial storage device.

Additionally, late in the initial deployment, the Earth-end of the tether is anticipated to experience excess forces over those of the away-end of the tether. Having a high inertial mass at GEO can act against a relatively short duration (a few hours) of imbalance between the earth-end of the tether and the above-GEO-end of the tether. Thus preventing a pull down as experienced in the Lang paper.

The relatively high inertial mass of the GEO ballast mass and low, but otherwise overwhelming force for the 20N thrusters, would result in slowly changing the orbital position of the GEO ballast mass as opposed to pulling down the tether. The change in position of the GEO ballast would remain within the design requirements of where the GEO position of the tether is permitted to wander. Without this reactionary mass, significant thrust capacity and fuel will be required to counteract the anticipated late stage forces.

VIII. Where to obtain the additional mass?

It is desirable that the extra mass not be dead-weight mass lifted from earth to GEO. This mass must serve additional useful purposes or must be readily available.

One potential source for readily available mass would be one or more decommissioned communication satellites.

GEO communication satellites, upon decommissioning, are customarily placed in a parking orbit about 300 km above GEO and then abandoned. Depending on the time chosen for deployment, one or more abandoned communications satellites might be in favorable position for capture and use as GEO ballast mass during initial deployment. It is unclear at this time if an abandoned communications satellite can claimed by anyone with the wherewithal to grab one similar to law-of-the-seas rules for shipwrecks. Perhaps by placing a few pounds of gravel (in a bag) into the path of the abandoned communications satellite and letting the satellite run into the gravel you could claim it ran aground, and thus claim salvage rights to the satellite.

A large abandoned communications satellite has a mass of around 2800 kg. Considering the cost/kg of lifting satellites into GEO one communications satellite might represent a savings of $60m of delivering dead-weight ballast to GEO.

A second class of mass for GEO ballast is that mass which is usable for other purposes:

- A new communications satellite co-launched and used as ballast prior to relocation and commissioning.
- A scientific satellite.
- Additional tether for later use in buildup of the space elevator.
- Additional equipment for later use in buildup of the space elevator.
- Additional fuel for later use in buildup of the space elevator.

IX. Targeting system

In addition to the GEO ballast a beacon satellite will be deployed at GEO and positioned to orbit slightly in front of the tether GEO ballast mass. This is depicted by the small free body to the right of the ballast mass in the figure depicting the Traditional, Bolo and Shoestring deployment techniques.

The purpose of the beacon satellite (which could be simulated in software) is to provide a targeting object for the control system. The beacon satellite (and deploying tether) will experience tidal forces, principally from the moon, sun, etc…. These tidal forces will cause deviations in the orbital positions of the beacon satellite and the deploying tether.

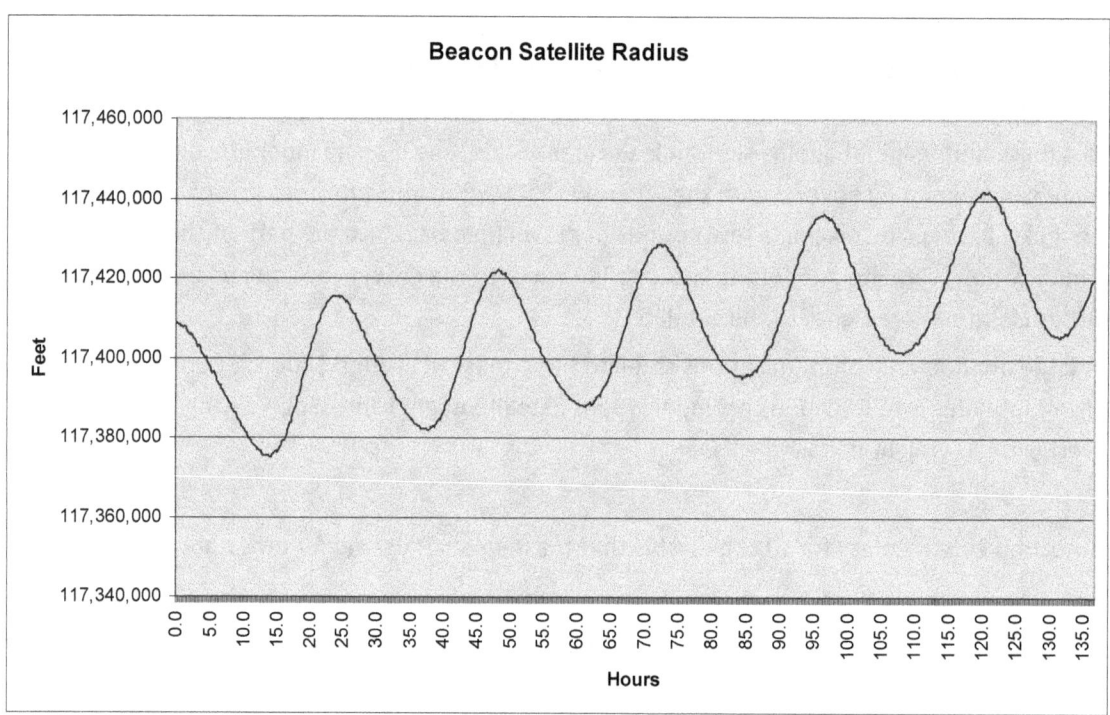

Figure 10

Lunar tidal effect is the higher frequency signal (~5.7 cycles), and the low frequency signal (~0.5 cycles) representing an initialization to a minor elliptical orbit.

The control force applied by the thrusters on the GEO ballast mass will target the tidally influenced beacon satellite orbit trail as opposed to a fixed position above earth. This is principally as a fuel saving procedure and a method to avoid undesired feedback through the un-damped spring nature of the tether.

X. Bolo Deployment Sequence

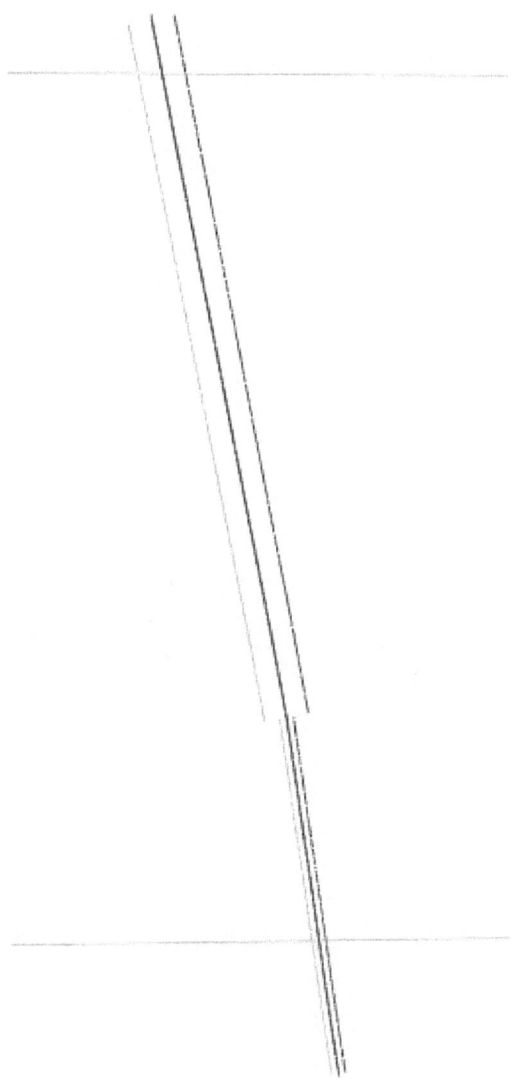

Bolo 27h 30m South View
Strain – Gold(left), Stress – Blue(right)
Figure 11

The deployment strategy chosen for this Bolo run is to set deployment rate such that the ratio of deployment rates of the above GEO portion of tether, to the below GEO portion of the tether, approximates the ratio of their remaining length. The top end spool contains approximately 2/3rds of the total tether and the bottom end spool contains the remaining portion of the tether. During the initial deployment phase, and using this deployment strategy, the 2:1 mass disparity between the top and bottom spools, the 2:1 offset difference from GEO, combined with the approximately complimentary slopes of the Ag + Ac curve at the spools, results in a tension differential across the constrained GEO position of the tether. Also of interest is the kink at the constrained GEO position (not visible in graph is GEO ballast mass with thrusters and fuel tanks).

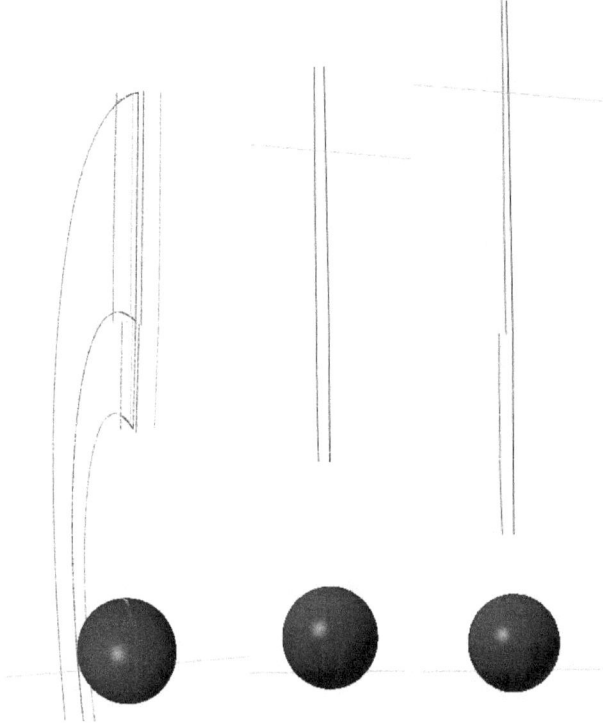

Bolo 196:35, 230:20, 260:10
View from East-East-Southeast
Tension – Red

Figure 12

The frames from left to right were taken at the times indicated in the caption above. At approximately 1/3rd through deployment (length), using rate proportional to remaining tether length, note that the tension disparity between the bottom and top portions of the tether across the constrained GEO ballast mass position (middle tracer line). Approximately 34 hours later the tension is in balance, and 30 additional hours the tension favors the bottom tether.

In the Bolo configuration, and using a deployment strategy of rate based on length remaining, we find that this places demands on the thrust capacity and fuel budget. Early in the deployment thrust is used to hold down the elevator while later on during deployment thrust is used to buoy the tether. A logical assumption is the thrust and fuel requirements for this deployment technique can be reduced by reducing the total mass of the initial deployed tether.

This investigator is exploring other means to keep the system in balance while reducing thrust and fuel requirement while increasing load capacity of the initial tether.

XI. Shoestring Deployment Sequence

One of the fuel conservative methods being explored by this investigator is the Shoestring deployment technique. In this technique the tether spools are held captive to the GEO ballast object. Instead of having the spools deploy outwards from GEO a loop of tether is deployed outwards. This technique constrains more mass near GEO for a longer duration during deployment.

The following figures are screenshots from the simulation program. Most views are primarily from a Southern vantage point.

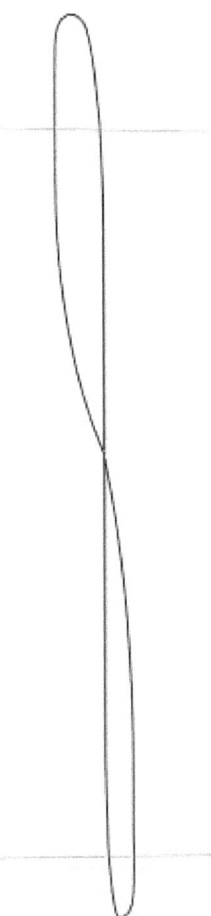

Shoestring 2h 20m bottom tether deploying at 6.3 fps

Figure 13

View is from the South, West to left, East to right. Position of intersection of tethers is at GEO. Ignore the green lines (artifact of screen capture).

Upper tether deployment rate slightly more than 2x bottom tether deployment rate. The deployment rates ratios approximate the remaining tether length ratios of each tether segment.

Note Coriolis effect: Bottom loop with Eastward drift, Top loop with Westward drift.

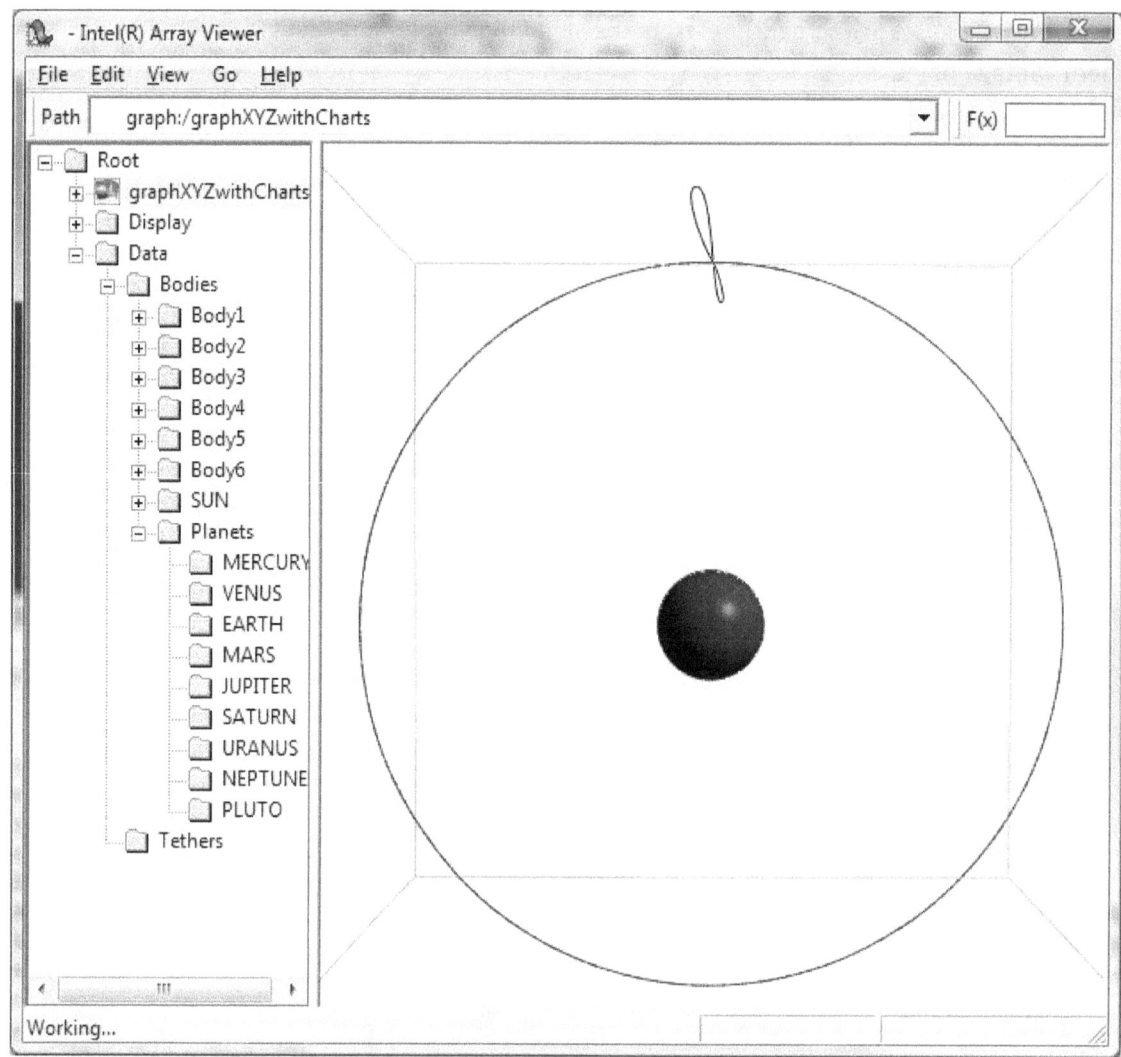

Shoestring 89h 40m Bottom Tether 413fps
Figure 14

This shot taken from due South includes Earth and tracer lines of the orbital paths of the GEO Ballast object, Beacon satellite, and deployment ends of tethers. Shoestring deployment holds spools near GEO ballast object during pay-out phase of tethers. The proximity of the two spools, to the GEO ballast object and beacon satellite combined with the scale of the graph results in what appears to be a single tracer line (there are four overlapping tracer lines).

On left side of image note browser tree showing in addition to Earth, the simulation model includes the tidal effects of the Moon, Sun and other major planetary bodies. Pluto, faithful dog as he is, is hanging around with his planet friends.

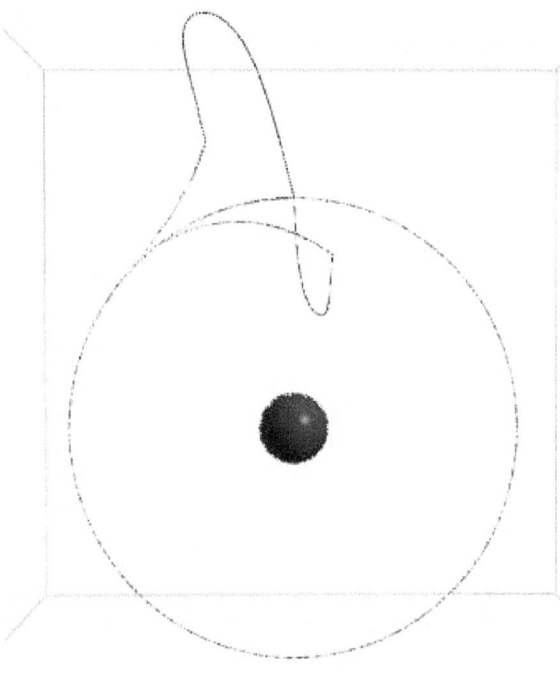

Shoestring 120h 20m

Figure 15

View from South, de-spooling of tethers occurred approximately 2 hours prior to screenshot (~10:00 clock hand for departure of trace lines). Purple lines are tracer lines, black lines are tethers.

C^LIMB Print readers, black and white lacking color, circle and Y branching from circle at 10:00 are purple.

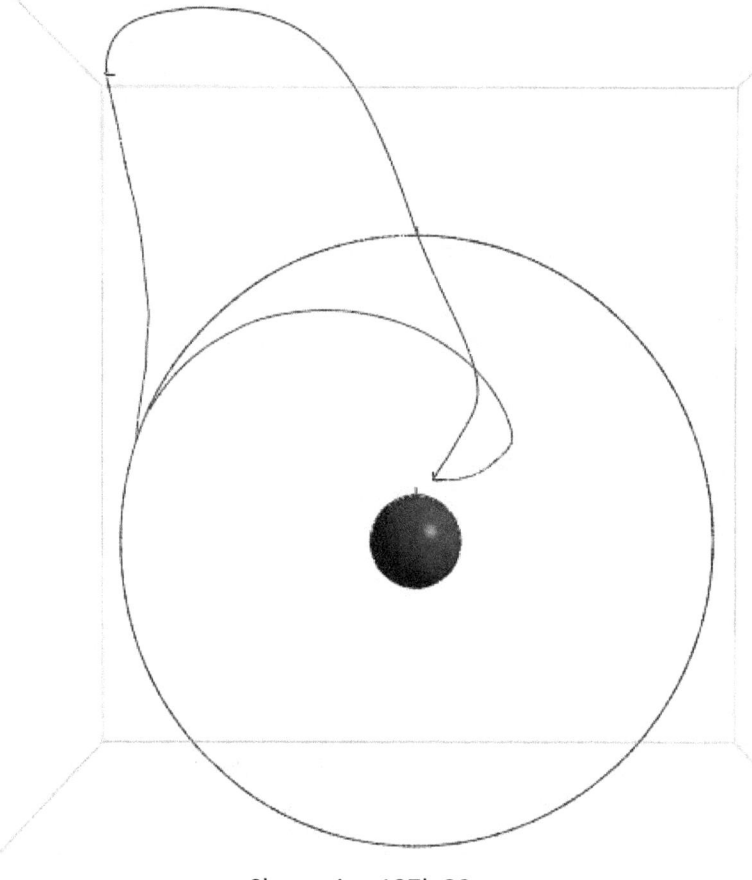

Shoestring 127h 20m
Figure 16

View from South, Earth-end of tether is plunging deeper into gravity well. Top-end tether position is lagging behind (not as vertically oriented as bottom tether), and indicates too late of departure from GEO ballast object. The result of failure to complete de-spooling of upper tether, earlier, results in tension deficit in top tether. This can be corrected by an earlier release of top tether. This will require an earlier start of deployment and/or faster deployment rate of top tether, or slower deployment rate of bottom tether. Finding the optimal deployment sequencing is a work in progress.

CLIMB Print readers, black and white lacking color, circle and Y branching from circle at 10:00 are purple.

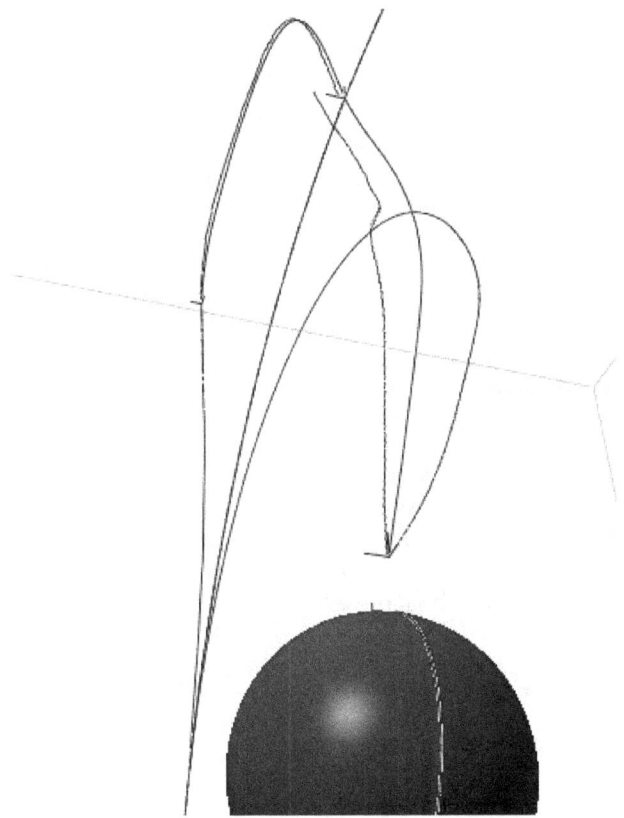

127h 10m view from below with tension charts.

Figure 17

Different view, same time, as prior screenshot. View from below and East of anchor point. To left of bottom tether (middle line as viewed) is a chart of the Tension. The charts in this simulation follow the tether. The charts are 3-dimensional lines running adjacent to the 3-dimensional tether line. Displacement of the tension line from tether line is the indication of the magnitude of the tension. In this screenshot Stress, Strain and Elastic Area are turned off and only Tension is shown. An interesting point to observe in this screenshot is as the bottom loop unfolds a fairly large tension spike occurs approximately half way up the tether as noted by the bulge to the left. When viewing the Strain data, excessive strain was found (~12%) along the tether at that position.

To help identify the ends of tethers (and anchor point on Earth) crosshairs enabled (orthogonal lines).

Tension also charted for top tether. Note tension disparity (distance from tether line) between bottom and top tethers bottom tether at GEO has approximately 5x the tension as the top tether. As stated with text accompanying prior screenshot, this tension disparity is primarily due to a sequencing error in the deployment solution. Sequencing the upper tether to release earlier would reduce or, ideally, eliminate the tension differential at GEO ballast object.

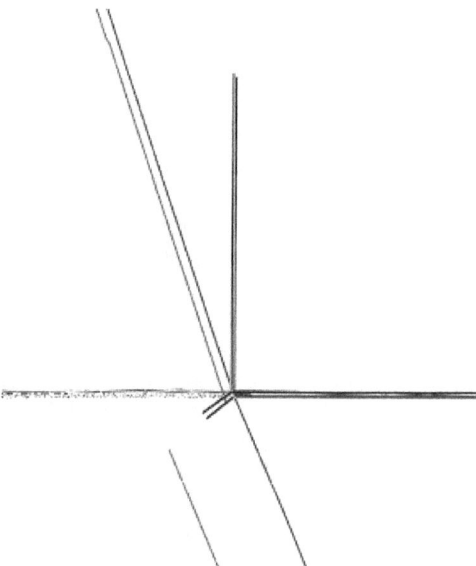

Shoestring 127h 10m view of GEO ballast object with beacon satellite and tension charts

Figure 18

Same time as prior two screenshots. View from South-South-East.

Note tension disparity between bottom tether and upper tether (red lines). Also note that the GEO ballast object orbital altitude is slightly below and in advance of that of the beacon satellite (upper orthogonal lines). This orbital departure is due to the tension differential exceeding the maximum thrust of the selected thrusters by approximately 3x the thrust capacity.

What's interesting about this screenshot is the inertial mass of the GEO ballast object is mitigating the control loss due to insufficient thrust. This illustrates that GEO ballast mass can be substituted for thrust capacity and fuel mass, at least for short durations.

	4	5	6	7
96	21.510040084291	8.256415750216e-009	2.605251568603e+009	0.095978750150
97	21.673292217469	8.256756161361e-009	2.624916104328e+009	0.096703045132
98	21.805006519370	8.257089655319e-009	2.640761627836e+009	0.097286452058
99	21.682998442224	8.257416232091e-009	2.625881732588e+009	0.096738309616
100	21.685371420927	8.257735891675e-009	2.626067448196e+009	0.096745061373
101	21.655149767600	8.258048634073e-009	2.622308335440e+009	0.096606836125

	4	5	6	7
96	3.828944989202	8.251589454194e-009	4.640251445442e+008	0.017095197652
97	3.490104943559	8.251589454194e-009	4.229615352209e+008	0.015581443568
98	3.845914700614	8.251589454194e-009	4.660816830458e+008	0.017173322180
99	4.517945825761	8.251589454194e-009	5.475243104181e+008	0.020170740545
100	3.888777468628	8.251589454194e-009	4.712761693022e+008	0.017359672706
101	3.959780712730	8.251589454194e-009	4.798809653233e+008	0.017680700874

Shoestring 127h 10m T1 and T2 data

Figure 19

The above two screenshots depict portions of the data tables for the states of Tether 1 (Earth-end) and Tether 2 (Away-end). From column 4 (left most) through column 7 is Tension, Elastic Area, Stress, and Strain. Rows 101 are the segments of the tether adjacent to the connection point on the tether to the GEO ballast object. This position on the tether will eventually become the GEO position of tether once the tether is stabilized.

The tension on the bottom tether (upper table) at GEO is ~21.66 lbf while the tension on the top tether (lower table) is ~3.96 lbf with a tension differential of 17.7 lbf. The maximum thrust of the selected thruster (CHT 20) is 5.53 lbf resulting in a system with a thrust deficiency of 12.17 lbf. Higher thrust capacity thrusters can be used to maintain altitude and position of Ballast object however in doing so removes an incentive to find a thrust minimum solution.

Tether 1 (top chart) which represents the Earth-end, higher tensioned tether, is experiencing a strain of ~9.74% in the vicinity of the GEO ballast object. This strain is above the desired working strain (4%) but below the maximum working strain (10%). When the actual tether material becomes available and the working properties are known, the tether profile can be revised to keep the maximum deployment strain below the maximum working strain.

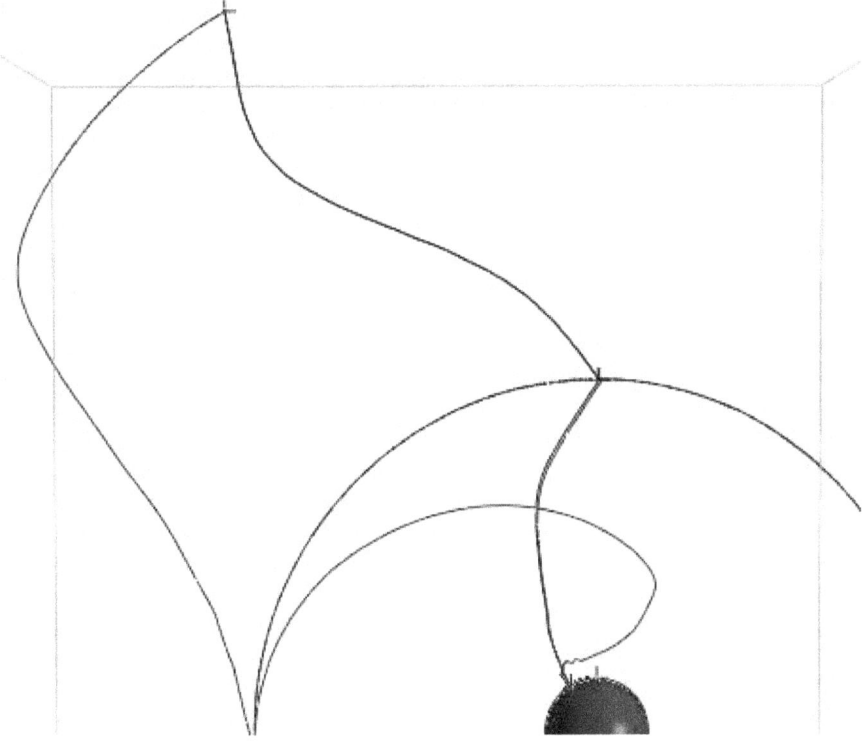

Shoestring 128h 40m (near) touchdown.

Figure 20

In this deployment simulation the intended results were not attained.

The landing point of the Earth-end of the tether misses the anchor point (note crosshairs on anchor point at 12:00 clock position on Earth). Approximate landing position ~2500 miles West of intended anchor point. It is evident that additional practice is required to come closer to the intended anchor point.

Also note, the top-end of the tether is too far from the zenith and thus has not reached the point of providing sufficient tension to support the tether. The principal corrective measure for this is to modify the deployment sequencing such that the top tether de-spools earlier and thus is given more time to approach near zenith. Additionally, simulations indicate that a thruster and fuel will be required on the top end of the tether to hasten the stabilization of the structure. The Earth-end of the tether being anchored (and whilst in atmosphere) requires no thrust for stabilization. If further simulations indicate some degree of control is required on entry by the Earth-end then it is anticipated that a low mass flight system be employed. Such as something similar to a radio controlled glider.

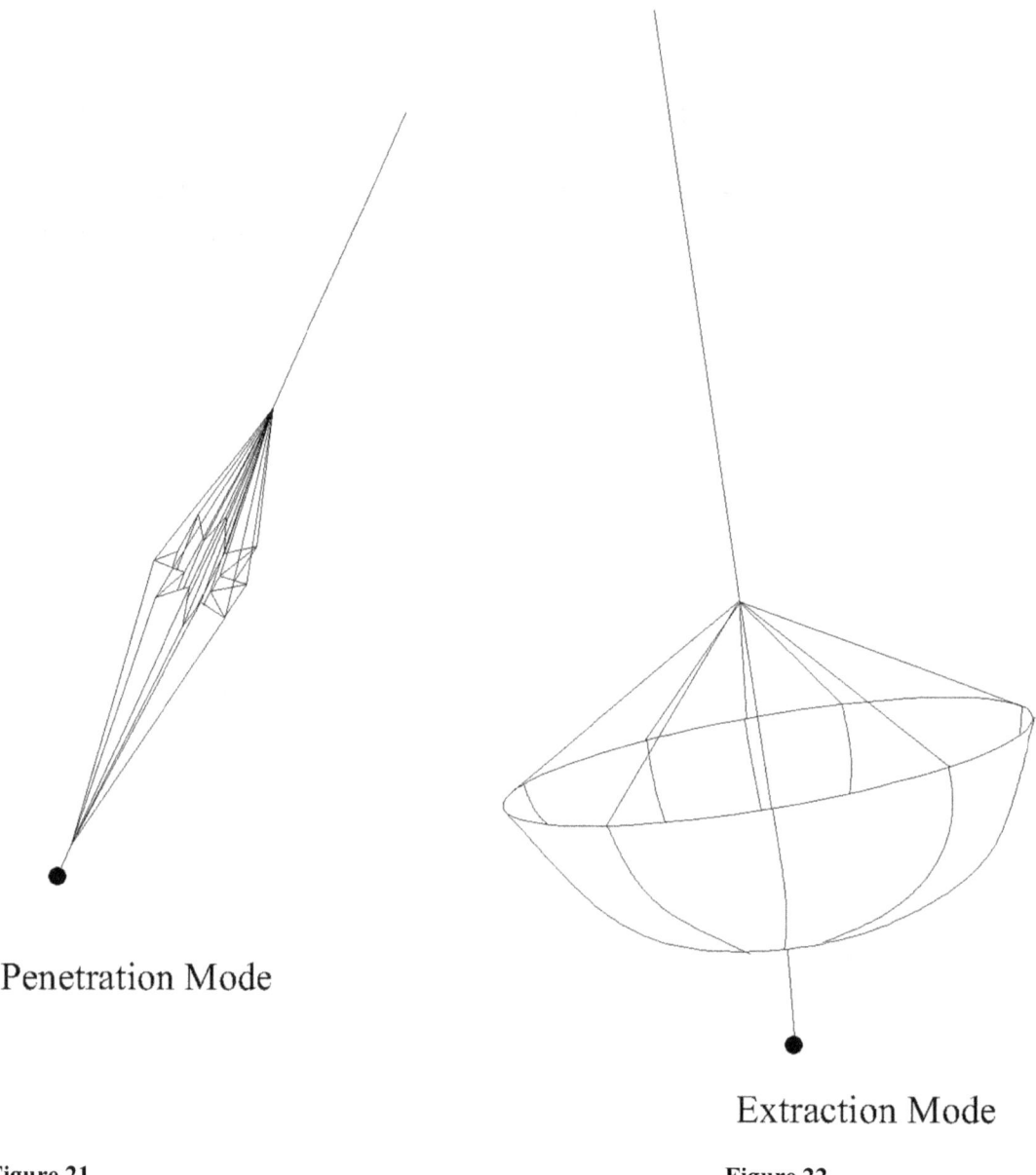

Penetration Mode   Extraction Mode

Figure 21   **Figure 22**

The Earth-end of the tether for the Shoelace configuration will experience two atmospheric modes: Penetration and Extraction. And two oceanic modes: Capture water ballast and float.

The tether will behave like a giant spring as well as like a giant pendulum. The very end of the Earth-end of the tether will contain a drogue chute and a small mass as depicted above. During Penetration Mode, the configuration of the small mass, drogue chute, and tether are such that the small mass pulls on the top of the drogue chute more so than the tether pulling on the shroud lines of the drogue chute, thus causing the small mass to lead the assemblage in penetration of the atmosphere. When the tether tension at the shroud lines exceeds that exerted by the gravitational force on the small mass through the small lanyard at the top of the drogue chute (bottom of diagram) the assemblage will attempt to be extracted from the atmosphere. This is called extraction mode. Extraction mode causes the drogue chute to fill with air. Penetration Mode has a small aerodynamic cross section whereas Extraction Mode has a high aerodynamic cross section.

The design criteria of the drogue chute has two additional features: it floats on water and it holds water as ballast to inhibit extraction after splashdown.

XII. Conclusion

Further studies are required for both the Bolo and Shoestring deployment techniques. These studies will refine:

- Tether profile
- Deployment sequencing
- Initial tether load capacity
- Ballast mass requirement
- Thruster and fuel requirements
- Duration of deployment
- Then subsequently layering up techniques

From these studies we can choose the deployment strategy with the highest probability of success. It is too early in the study phase to determine which is the better strategy. The shoestring appears to be more fuel conservative; however, all avenues of equilibrium assistance have yet to be studied. The Bolo configuration appears to be less of a "wild ride" than the Shoestring.

For further information, direct inquiries to:

Jim Dempsey
85 Cove Lane
Oshkosh, WI 54902
Jim.00.dempsey@gmail.com

Mr. Dempsey has been performing a full-time, self-funded, research project going on 10 years. He holds two U.S. Patents relating to Space Elevator design and deployment techniques.

Acknowledgments

I wish the thank David D. Lang Associates, Seattle, WA and in particular Dr. Lang for the countless number of hours he and his organization spent in developing GTOSS. GTOSS was the basis of the simulation code used for my studies. The original code went through extensive changes for this author's purposes, principally in the areas of feature and performance enhancement. Dr. Lang has expressed his concern that having not personally run verification test on the changes made by this author, that he wishes to disclaim any certification as to the accuracy of the modified code. Regardless of this certification, this author is greatly indebted to the excellent work Dr. Lang and his associates performed in producing the original GTOSS (tether simulation code).

References

Patents

James G. Dempsey, Oshkosh, WI, U.S. Patent "System and method for space elevator", 6,981,674, January 3, 2006

James G. Dempsey, Oshkosh, WI, U.S. Patent "System and method for space elevator deployment", 7,971,830, July 5, 2011

TRANSVERSE VIBRATION OF THE SPACE ELEVATOR TETHER WITH SPACEPORTS

S.A.Ambartsumian, M.V.Belubekyan, K.B.Ghazaryan, R.A.Ghazaryan
Institute of Mechanics of National Academy of Sciences, Yerevan, Armenia

Abstract: Conditioned by technical requirement the space elevator structure may have several space spaceports positioned along length of the tether. In the paper a new problem is proposed for dynamic behavior of the space elevator structure where the tether is considered as a beam made from different flexible materials, meanwhile spaceports are considered as attached counterweight masses positioned on different orbit levels including the end of the tether. By means of variation methods the appropriate eigenvalue boundary problem is established, the solution thereof determines the structure frequencies with a great accuracy. A detailed numerical analysis is carried out describing the effects of tether material non-homogeneity and attached masses on the dynamic's characteristics of tether.

Nomenclature

F_1 = gravity inward specific force

F_2 = the Earth centrifugal outward specific force

γ = coordinate along tether length

ρ = bulk density of a tether material

R_0 = the Earth equatorial radius (6378 km)

g_0 = gravity force acceleration on the Earth surface ($9.806 m\,\text{sec}^{-2}$)

g = dimensionless function of resulting gravity force acceleration

ω_E = rotational angular velocity of the Earth ($7.29\,10^{-5}\,\text{sec}^{-1}$)

σ = elastic stress

x_0 = point of the Earth geostationary orbit (35786 km)

L = tether dimensionless length

L_0 = "limiting length" of tensile tether

E = elasticity module of tether material

m_0 = mass of the spaceport attached at point x_0

m_L	=	mass of the spaceport attached at the tether's end point
m	=	mass of a tether
S	=	cross-sectional area of the tether
J	=	cross-sectional bending moment of inertia of tether material
ω	=	tether transversal vibration physical frequency
Ω	=	tether transversal vibration dimensionless frequency
U	=	longitudinal displacement of tether points
W	=	transversal displacement of tether points

1. Introduction

The mathematical concept of the space elevator was first introduced in Ref.1, where problems of buckling, strength and vibration were discussed. An overview, developments of the space elevator concept and detailed design for construction and operation are, particularly, in Ref.2, Ref.3, Ref.4. The concept of the space elevator refers to a structure which reaches from the surface of the Earth equator to geostationary orbit and a counterweight beyond the level extended out to 144,000 km. This structure would be held in tension (conditioned by resulting force of the Earth gravity inward force defined by the Newton gravity law and centrifugal outward force caused by the Earth's daily spinning) between the Earth and the counterweight in which the maximum tensile strength would reach up to 120 GPa. Besides huge tensile stresses the structure undergoes large deformations also. Materials currently available do not meet these requirements although advances in carbon nanotube technology could make it possible to create a strong type of carbon nanotube materials with Young's moduli as high as 1500GPa and tensile strengths of up to 200 GPa Ref.5, Ref.6..

The space elevator stationary structure makes it possible to place permanent spaceports on it for the deployment of scientific and technological activities in the space, greatly easing the logistic problems, and replacements of personnel are compared with those used to date orbiting space stations. These spaceports will be the starting point in resolving such global issues of the use of terrestrial resources and needs of near and far space.

Here transversal vibration problems are discussed for a space elevator structure where the spaceports are considered as attached masses.

2. Problem formulation

Let us consider a very long elastic tether anchored on the Earth equatorial point. The tether subject to the action of Earth gravity inward specific force F_1, defined by Newton gravity law and centrifugal outward specific force F_2 caused by the Earth daily spinning

$$F_1(\gamma) = -\frac{\rho(\gamma) g_0 R_0^2}{(R_0 + \gamma)^2}; \quad F_2(\gamma) = \rho(\gamma) \omega_E^2 (R_0 + \gamma) \qquad 0 \leq \gamma \leq l \qquad (1)$$

Introducing the dimensionless coordinate $x = \gamma/R$ and taking into account that

$$\frac{\omega_E^2 R_0}{g_0} \approx \frac{1}{288} \qquad (2)$$

we can write the resulting specific force $F = F_1 + F_2$ as

$$F(x) = \rho(x) g_0 g(x) \qquad (3)$$

where

$$g(x) = \left[\frac{(1+x)}{288} - \frac{1}{(1+x)^2}\right] \qquad (4)$$

is a dimensionless resulting gravity force acceleration at tether point x

When the cross-section area S of the tether is a constant, the equation determining the elastic stress $\sigma(x)$ can be written as

$$\frac{d\sigma(x)}{dx} + R\rho(x) g_0 g(x) = 0 \qquad (5)$$

We shall model the tether's spaceports as attached masses positioned along the length of the tether.

At point x_1 of attached mass position stress jump occurs defined by formulae

$$[\sigma]_{x=x_1} = \frac{m_1 g_0}{S} g(x_1) \qquad (6)$$

In the Earth geostationary orbit point x_0 $[\sigma]_{x=x_0} = 0$, since the resulting gravity force acceleration at this point is equal to zero.

If there is an attached mass at the tether's end point, then instead of Eq. (6) we have

$$\sigma(L) = \frac{m_L g_0}{S} g(L) \qquad (7)$$

Let us now consider the compound tether made from two different materials separated by attached mass m_0 positioned at $x = x_0$ and having attached mass m_L positioned at the tether's end $x = L$

For the compound tether taking into account that

$$S = \frac{m}{R[\rho_1 x_0 + \rho_2(L - x_0)]} \qquad (8)$$

Equation (7) can be rewritten as

$$\sigma(L) = Rg_0\rho_1\beta_L g(L) \tag{9}$$

$$\beta_L = \frac{m_L\left[x_0 + \theta(L-x_0)\right]}{m}; \quad \theta = \frac{\rho_2}{\rho_1}; \tag{10}$$

Solving Eq. (5) under condition of stress continuity at $x = x_0$ we come to the following solution for stress functions

$$\begin{aligned}\sigma_1(x) &= Rg_0\rho_1\sigma_{10}(x) & 0 \leq x \leq x_0 \\ \sigma_2(x) &= Rg_0\rho_1\sigma_{20}(x) & x_0 \leq x \leq L\end{aligned} \tag{11}$$

$$\begin{aligned}\sigma_{10}(x) &= f(x) - f(x_0) + \theta\left[f(x_0) - f(L)\right] + \beta_L g(L) \\ \sigma_{20}(x) &= \theta\left[f(x) - f(L)\right] + \beta_L g(L)\end{aligned} \tag{12}$$

$$f(x) = \frac{x}{1+x} - \frac{x(2+x)}{576}$$

Since function $\sigma(x)$ reaches its maximum at $x = x_0$ and could have only one zero in the interval $0 \leq x \leq x_0$, the condition $\sigma_1(0) = 0$ ensures $\sigma(x) \geq 0$ at any point along the tether.

Condition $\sigma_1(0) = 0$ can be written as

$$\theta(f(x_0) - f(L)) - f(x_0) + \beta_L g(L) = 0 \tag{13}$$

Equation (13) defines the minimum "limiting length" L_0 below which compression stresses may arise near tether base which can make the tether unstable Ref.(3).

Using Hooke's elasticity law

$$\sigma(x) = E\frac{dU(x)}{dx} \tag{14}$$

the tether longitudinal displacements with fixed base $U(0) = 0$ can be found as

$$\begin{aligned}U_1(x) &= \frac{R^2 g_0\rho_1}{E_1} s(x) \\ U_2(x) &= \frac{R^2 g_0\rho_1}{E_2}\left[\theta(s(x) - s(x_0)) + (\beta_L g(L) - \theta f(L))(x-x_0)\right] + \frac{R^2 g_0\rho_1}{E_1} s(x_0)\end{aligned} \tag{15}$$

where

$$s(x) = x\left(1 - \frac{3x + x^2}{1728}\right) - \log(1+x) \tag{16}$$

Since the stress function has zero at tether base $x = 0$, (when mass $m_L = 0$ the stress function has zero also at tether's end $x = L$) in order to avoid the consideration of singular differential equation of second-order we shall study transversal vibration of stressed tether modeling tether as an elastic beam (beam model) Representing the transverse displacements as

$$W_1 = W_{10}(x)\exp(i\omega t) \quad W_2 = W_{20}(x)\exp(i\omega t) \tag{17}$$

and introducing dimensionless parameters

$$A_1 = \frac{E_1 J}{\rho R g_0 S}; \quad A_2 = \frac{E_2 J}{\rho R g_0 S} \quad \Omega = \omega\sqrt{\frac{R}{g_0}} \tag{18}$$

we come to the following governing equation of transversal bending vibration of stressed tether

$$\begin{aligned} A_1 \frac{\partial^4 W_1}{\partial x^4} - \frac{\partial}{\partial x}\left(\sigma_{10}(x)\frac{\partial W_1}{\partial x}\right) - \Omega^2 W &= 0 \\ A_2 \frac{\partial^4 W_1}{\partial x^4} - \frac{\partial}{\partial x}\left(\sigma_{20}(x)\frac{\partial W_1}{\partial x}\right) - \Omega^2 W &= 0 \end{aligned} \tag{19}$$

This equation must be considered with the following boundary and contact conditions

$$W_{10}(0) = 0 \quad \left.\frac{\partial W_{10}}{\partial x}\right|_{x=0} = 0 \quad \left.\frac{\partial^2 W_{20}}{\partial x^2}\right|_{x=L} = 0 \quad \left.\left(A_2 \frac{\partial^3 W_{20}}{\partial x^3} + \sigma_{20}\frac{\partial W_{20}}{\partial x}\right)\right|_{x=L} = 0 \tag{20}$$

$$W_{10}(x_0) = W_{10}(x_0) \quad \left.\left(\frac{\partial W_{10}}{\partial x} - \frac{\partial W_{20}}{\partial x}\right)\right|_{x=x_0} = 0 \quad \left.\left(E_1 \frac{\partial^2 W_{10}}{\partial x^2} - E_2 \frac{\partial^2 W_{20}}{\partial x^2}\right)\right|_{x=x_0} = 0$$

$$\left.\left(A_1 \frac{\partial^3 W_{10}}{\partial x^3} + \sigma_{10}\frac{\partial W_{10}}{\partial x} - A_2 \frac{\partial^3 W_{20}}{\partial x^3} - \sigma_{20}\frac{\partial W_{10}}{\partial x} + \beta_0 \Omega^2 W_{10}\right)\right|_{x=x_0} = 0 \tag{21}$$

Here

$$\beta_0 = \frac{m_0\left[x_0 + \theta(L - x_0)\right]}{m} \tag{22}$$

The problem under consideration can be generalized for a tether with several attached masses positioned both the higher and lower levels from the Earth geostationary orbit. But we here confine ourselves to the solution of the problem with two attached masses aiming to propose method of solutions of such problems. Indeed the vibration problem of tether with more than two spaceports is very interesting and must be considered in the future.

3. Solution of the model problem

The solution of this boundary value problem experiences some mathematical complexities. We will propose a more simple approach to this problem based on the solution of stated below modeling problem.

First let us consider the homogeneous tether without attached masses. In this case the governing equation and boundary conditions can be written as

$$A\frac{d^4W}{dx^4} - \frac{d}{dx}\left(f(x)\frac{dW}{dx}\right) - \Omega^2 W = 0 \tag{23}$$

$$W(0) = 0 \quad \left.\frac{\partial W}{\partial x}\right|_{x=0} = 0 \quad \left.\frac{\partial^2 W}{\partial x^2}\right|_{x=L} = 0 \quad \left.\frac{\partial^3 W}{\partial x^3}\right|_{x=L} = 0 \tag{24}$$

We expand the solution of Eq.(23) into series with respect to the eigenfunctions of Sturm–Liouville problem of a vibrating cantilever beam

$$W(x) = \sum_j w_i^0 w_j(x) \tag{25}$$

where $w_j(x)$ are the normalized orthogonal functions satisfying boundary conditions Eq(24)

$$w_j(x) = \frac{F(p_j)(\sinh(p_j x) - \sin(p_j x)) + \cos(p_j x) - \cosh(p_j x)}{\sqrt{L}} \tag{26}$$

$$F(p_j) = \frac{\cos(p_j L) + \cosh(p_j L)}{\sin(p_j L) + \sinh(p_j L)} \tag{27}$$

$$\int_0^L w_j w_i \, dx = \delta_{ij}$$

The eigenvalues $p_j L$ are the roots of the following equation

$$\cos(p_m L)\cosh(p_m L) = -1 \tag{28}$$

The first four roots of Eq.( ) are

$$p_1 L = 1.8751, \quad p_2 L = 4.6940 \quad p_3 L = 7.8547 \quad p_4 L = 10.9955$$

Substituting Eq.(25) into Eq. (23) after applying the Galerkin orthogonalization technique leads to matrix equation

$$K_{ij} w_i^0 = \Omega^2 \delta_{ij} w_i^0 \qquad (29)$$

where

$$K_{ij} = \int_0^L f(x) \frac{\partial w_j}{\partial x} \frac{\partial w_i}{\partial x} dx + \delta_{ij} \left(A p_j^4 + \int_0^L f(x) \frac{\partial w_j}{\partial x} \frac{\partial w_i}{\partial x} dx \right) \qquad (30)$$

Thus, the vibration problem results in determination Ω_j^2 as eigenvalues of the symmetric positive defined matrix K.

4. Numerical results

Let us consider the following numerical test example for a tether made from carbon nanotube composite: the cross section of the tether is an elongated rectangular with sides a, b; $J = ba^3/12$, $S = ab$, $a = 1.0 m$, $E = 4.843 \cdot 10^{13}$ Pa, $\rho = 5000 \cdot kg \cdot m^{-3}$ kg, $A = 3.146 \cdot 10^{-13}$, $L_0 = 22.5052$.

Determining eigenvalues of matrix K we have

$$\Omega_1 = 0.05859, \quad \Omega_2 = 0.14478, \quad \Omega_2 = 0.2418, \quad \Omega_4 = 0.347205 \qquad (31)$$

Let us note that the natural frequencies of the beam are

$$\Omega_{bj} = \sqrt{A} p_j^2 \quad \Omega_{b1} = 1.962 \cdot 10^{-6}, \quad \Omega_{b2} = 1.232 \cdot 10^{-5};$$
$$\Omega_{b3} = 3.456 \cdot 10^{-5}, \quad \Omega_{b4} = 6.767 \cdot 10^{-5}$$

In addition to afore mentioned calculus we can note that if instead of Eq. (26) we take we take the following orthogonal and normalized polynomial functions

$$u_1(x) = \frac{3\sqrt{5} x^2 (6L^2 - 4Lx + x^2)}{2L^4 \sqrt{26L}};$$
$$u_2(x) = \frac{\sqrt{165}}{2L^5 \sqrt{4238L}} x^2 (326 L^3 - 824 L^2 x + 661 L x^2 - 182 x^3) \qquad (32)$$

satisfying the boundary conditions of cantilever beam, we come to the following values of Ω, which are very close to previous values

$$\Omega_1 = 0.060079, \qquad \Omega_2 = 0.14604$$

In Table 1. the data are presented for four physical frequencies ω_j and periods $T_j = 2\pi / \omega_j$.

$\omega_j(s^{-1})$	$7.262 \cdot 10^{-5}$	$17.96 \cdot 10^{-5}$	$29.982 \cdot 10^{-5}$	$43.05 \cdot 10^{-5}$
$T_j(s)$	86484	34990	20952	14590
$T_j(h)$	24.023	9.719	5.820	4.052

Table 1. Frequencies and periods of the tether vibrations

Let us note that the first frequency of the tether is very close to the rotational angular velocity of the Earth. This fact can be interpreted as no more than an interesting coincidence.

Since the tether dimensionless bending rigidness A is rather small, we will consider now the other boundary value problem where the tether is modeled as a stretched string, which can be named as 1D membrane model.

$$\tilde{\sigma}\frac{\partial^2 W}{\partial x^2} + \Omega^2 W = 0$$

$$W(0) = 0; \qquad \left.\frac{\partial W}{\partial x}\right|_{x=L} = 0 \tag{33}$$

Here $\tilde{\sigma}$ is a constant which would be defined as follows.

The eigenvalues of Eq.(33) can be found as

$$\Omega_{0j} = \frac{\pi\sqrt{\tilde{\sigma}}}{2L}(2j-1)]; \ j = 1, 2... \tag{34}$$

Equating the first frequency $\Omega_{0j}(j=1)$ with $\Omega_1 = 0.05859$ we come to $\tilde{\sigma} = 0.70469$ which is approximately equal to 91% of the maximum stress value $\sigma_{10}(x_0)$. The other frequencies of Eq.(34) corresponding to $\tilde{\sigma} = 0.70469$ are $\Omega_{02} = 0.1757$, $\Omega_{03} = 0.2929$, $\Omega_{04} = 0.4101$. Let us note that these values average 20% overestimate the bending frequencies defined by Eq.(31).

Finally based on the derived results instead of Eq. (19-21) we shall consider the vibration problem of compound tether with two attached masses by means of the 1D membrane model boundary value problem

$$\tilde{\sigma}_0 \frac{\partial^2 W_1}{\partial x^2} + \Omega^2 W_1 = 0 \qquad 0 < x < x_0$$

$$\tilde{\sigma}_0 \frac{\partial^2 W_2}{\partial x^2} + \frac{\rho_2}{\rho_1}\Omega^2 W_2 = 0 \qquad x_0 < x < L \tag{35}$$

Boundary and contact conditions

$$W_1(0) = 0; \quad \left(\tilde{\sigma}_0 \frac{\partial W_2}{\partial x} + \beta_L \Omega^2 W_2\right)\bigg|_{x=L} = 0 \qquad (36)$$

$$W_1(x_0) = W_2(x_0);$$

$$\left(\tilde{\sigma}_0 \frac{\partial W_1}{\partial x} - \tilde{\sigma}_0 \frac{\partial W_2}{\partial x} + \beta_0 \tilde{\Omega}^2 W_1\right)\bigg|_{x=x_0} = 0$$

Length L is determined from Eq.(13).

Representing solutions as

$$W_1(x) = C_1 \sin(qx) + C_2 \cos(qx);$$
$$W_2(x) = C_3 \sin(rx) + C_4 \cos(rx); \qquad (37)$$

$$r = \tilde{\Omega}\sqrt{\theta \tilde{\sigma}_0^{-1}} \quad q = \tilde{\Omega}\sqrt{\tilde{\sigma}_0^{-1}} \qquad (38)$$

Satisfying Eq.(34-36) results in the following equation determining eigenvalues $\tilde{\Omega}$

$$\sqrt{\theta \tilde{\sigma}_0}\left[\tilde{\sigma}_0 + \tilde{\Omega}(\beta_0 + \beta_L)\sqrt{\tilde{\sigma}_0}\tan(qx_0)\right] = $$
$$= \left[(\theta\tilde{\sigma}_0 - \beta_0\beta_L\Omega^2)\tan(qx_0) - \beta_L\tilde{\Omega}\sqrt{\tilde{\sigma}_0}\right]\tan(Lr - rx_0) \qquad (39)$$

β_L	$\Omega_1 = \tilde{\Omega}_1/\Omega_{01}$	L_0
0	1	22.52
0.1	1.10	20.54
0.2	1.19	20.82
0.3	1.28	17.81
0.5	1.37	16.81
0.7	1.59	14.62
1	1.78	13.17
Table2. Frequency and limiting length data		

Based on the solutions of Eq.(39) in Tables.2-4 the values of the first frequency (normalized to the frequency of homogeneous tether without masses) are presented depending on non-homogeneity parameter θ and values of relative masses.

Data Table2. show that when $\beta_0 = 0; \theta = 1$ minimum frequency and "limiting length" increase with the increase of value of mass attached on the tether end, On the other hand the numerical analysis of Eq.(39) shows that minimum frequency practically does not depend on β_0 when $\beta_L = 0; \theta = 1$. In this case the "limiting length" is a constant one equal to $L_0 = 22.52$.

In Table3. the data are presented for minimum frequency and "limiting length" depending on the non-homogeneity parameter θ and value of attached mass β $(\beta_L = \beta_0 = \beta)$. In Table4. the minimum frequency is given related to the non-homogeneity parameter when $\beta_L = \beta_0 = 0$.

θ	β	$\Omega_1 = \tilde{\Omega}_1/\Omega_{01}$	L_0	θ	$\Omega_1 = \tilde{\Omega}_1/\Omega_{01}$	L_0
1.0	0	1	22.52	1.0	1	22.52
1.1	0.1	1,11	19.64	1.1	0.993	21.62
1.2	0.2	1.21	17.28	1.2	0.986	20.82
1.5	0.3	1.330	14.38	1.5	0.971	18.98
2	0.5	1.565	10.93	2	0.952	16.93
3	0.7	1.815	8.57	3	0.91	14.57
5	1	2.138	6.96	5	0.86	12.29

Tables 3. 4. Frequency and limiting length data (additional)

On Fig.1 graphs show the dependence of the longitudinal displacement function $u_0(\theta) = U_2(L)/U_2(x_0)$ related to coefficient θ, for several values of parameter $k = E_1/E_2$ under $\beta_L = 0.15$. The lower graph corresponds to $k = 1$, middle graph corresponds to $k = 0.6$, upper graph corresponds to $k = 0.2$. Numerical analysis showed that the influence of β_0 on displacement function is not significant. The displacement of point of the Earth geostationary orbit is constant and does not depend of masses values β_L, β_0 and parameters θ, k.

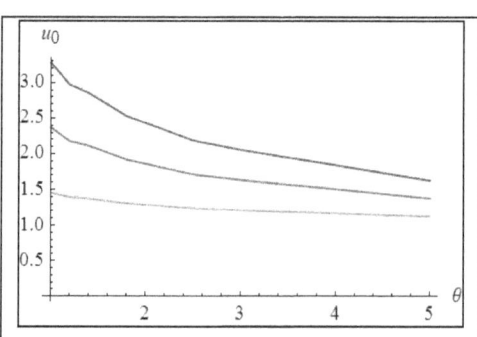

Fig 1. Dimensionless displacement function related to parameter θ

For a homogeneous tether $k = 1$ made from carbon nanotube composite with $E = 4.843$ TPA, $\rho = 5000 \cdot kg \cdot m^{-3}$ absolute elongation of x_0 point of the Earth geostationary orbit is equal to 444.5 km.

In addition we note that we did not present the numerical results for compound tether with $E_1 > E_2, \rho_1 > \rho_2$ since in these cases the compound tether is not optimal. In these cases the values of maximal stress, longitudinal displacement at the tether's end and "limiting length" exceed the corresponding values of homogeneous tether.

5. Conclusions

Considering the tether as a flexible beam without spaceports the first four natural frequencies of transversal vibration are determined with great accuracy. It is noted that the first frequency of the tether practically coincides with the rotational angular velocity of the Earth. Assuming the tether's spaceports as attached masses the vibration problem is generalized to the case of a compound tether made from two different materials. Within the framework of the suggested 1D membrane model the minimum frequencies are determined related to non-homogeneity parameters and masses attached at GEO point and tether's end. Numerical results are presented for dimensionless longitudinal displacement function defining elongation at the tether's end.

6. References

[1]   J. Pearson. The orbital tower: a spacecraft launcher using the Earth's rotational energy, Acta Astronautica., 1975, Vol. 2. pp. 785-799.

[2]   B. C Edwards, Design and Deployment of a Space Elevator. 2000., Acta Astronautica. 47, pp.735-744.

[3]   S..A.Ambartsumian, M.V.Belubekyan, K.B.Ghazaryan, Stability of superconducting cable used for transportation of electrical current from space, Acta Astronautica, Vol: 66, 2010, pp. 563-566.

[4]   S.A.Ambartsumian, M.V.Belubekyan, K.B.Ghazaryan, Optimal design of the space elevator tether, Climb, v.1,no.3,2011,p.11-23.

[5]   P. J. F. Harris, Carbon nanotube composites, Journal of International Materials Reviews, v 49, no 1, 2004

[6]   Brambille G. An updated review of nanotechnologies for the space elevator tether, Climb, v.1,no.3,2011,p.3-11

SPACE ELEVATOR POWER SYSTEM ANALYSIS AND OPTIMIZATION

Ben Shelef

the Spaceward Foundation

Abstract: This paper lays out the basics constraints for a Space Elevator power system, performs parameter optimization, and compares the results with real-life technology parameters. The paper also considers the special case of solar climbers that have the additional constraint of a once-per-day launch rate.

Motivation

Being a transportation system, a primary goal in designing a Space Elevator system is maximizing payload throughput. Typical design parameters that can be varied to achieve this maximum are: Payload fraction (If we allocate more mass to the power system, a climber can move faster, but can carry less payload), Frequency of launch and number of simultaneous climbers on the ribbon, etc.

Since the Space Elevator is linearly scalable, we normalize the calculations by the maximum mass that is allowed to hang from the bottom of the tether (m_{max}). Thus a "20-ton" Elevator is one that can support a single 20 ton climber at ground level, including all safety margins. (In turn, m_{max} is a fraction of the total tether mass m_T, with the tether mass fraction TMR defined as m_{max}/m_T)

So for example, for a 20 ton elevator, we might choose to use 15 ton climbers (0.75 M_{max}), and the tether might weigh (depending on the nanotubes) anywhere between 200 and 400 m_{max}. We define a "Standard throughput unit" (STU) as being able to launch one m_{max} per year. By default, throughput will refer to the amount of payload carried.

To simplify matters, we divide the mass of the climber into payload and power system, assuming the "dead structure" is light in comparison to either of them. (The power system mass includes all components that scale with the power level, such as PV panels, motors, power electronics and radiators.

Ascent power profile

The power required to move a mass at a certain velocity is a function of the effective gravity at the altitude that the climber is at: $P = m \cdot g_{eff} \cdot v$, where $g_{eff}(r) = g(r_e/r)^2 - \omega^2 r m$ (r_e=6400 km is the radius of the earth, and ω_e = 7.3E-5 rad/sec, the angular velocity of the Earth and g is the surface gravity).

For $r < 2r_e$, we can approximate this very well as simply $g_{eff}(r) = g(r_e/r)^2$, which means that for a specific power system, the climber's velocity will increase according to a square law $v(r) = P/(m \cdot g) \cdot (r/r_e)^2$ as it moves out, until it reaches some maximum terminal velocity v_T determined by the tether handling system. From that point onwards, the climber moves at v_T, and the power system is under-utilized.

Defining power density ρ_P as power system mass divided by the power it can deliver, and power system mass fraction β as the mass of the power system divided by the mass of the climber, the total available power is $P = m \cdot \beta \cdot \rho_P$, the initial velocity is $v_e = \beta \cdot \rho_P/g$, and so the velocity is $v(r) = \min[V_T, v_e \cdot (r/r_e)^2]$. The terminal velocity point r_T is located where the climber reaches v_T, which happens at $v_T = v_e \cdot (r_T/r_e)^2$, or at $r_T = r_e(v_T/v_E)^{0.5}$. The amount of payload carried by the climber is of course $(1-\beta)m$.

The formula for the time it takes a climber that is following a constant power velocity profile to cover the distance between r_e and r (for $r < r_T$) is:

$$t = \int_{x=r_e}^{x=r} dx / v(x) = \int_{x=r_e}^{x=r} v_e^{-1}(x/r_e)^{-2} dx = (r_e/r)(r-r_e)/v_e = t_0/Q \tag{1}$$

Where $t_0 = (r-r_e)/v_e$ is the time it would have taken the climber to cover the distance if it were moving at a constant velocity, and $Q = r/r_e$ is the radius ratio.

The distance traveled by a constant power velocity climber (relative to r_e) is: $d/r_e = (r-r_e)/r_e = t/(r_e/v_e - t)$.

Handoff

The spacing between climbers can be characterized by a handoff fraction k_H, such that a new climber is launched when the old climber reaches the altitude where $g_{eff}/g = k_H$. The handoff altitude $r_H = r_e/k_H^{0.5}$ is the location where this happens, and the time since launch that this happens at is defined as t_H)

To keep the total tether load from exceeding the equivalent of one m_{max} climber at ground level, The mass of each climber can only be $m = (1-k_H)m_{max}$, so that the geometrical series $(1-k_H)+(1-k_H)k_H+(1-k_H)k_H^2... = 1$. (This is slightly conservative, since the spacing between the climbers does not remain constant, and there are only a finite number of climbers).

If the initial parameters are such that $r_T > r_H$, (or $(v_T/v_e)^{0.5} > k_H^{-0.5}$) then the climber will follow a constant-power velocity profile all the way out to the handoff point. We call this a "power limited" profile. Otherwise, the climber will "max-out" on the way to the handoff point, and the profile is called "speed limited". There is also the hypothetical possibility the $r_T < r_e$, which means that the climber power system starts out under-utilized, even at ground level. This is clearly a non-optimal case.

Throughput

The payload per climber is therefore $m_P = (1-\beta)(1-k_H)$ and the mass throughput of the system is $PMT = (1-\beta)(1-k_H)/t_H$. The more frequently we launch climbers, the smaller each one can be, but the larger the throughput. This trend continues to the limiting case of a continuous (variable speed) belt of cargo, though we see no practical way of implementing this case. Similarly, the faster we move, the more climbers we can launch, but the larger power systems leave less room for payload.

The parameters dictated by technology are the power density ρ_P and the terminal velocity v_T. The variables we can tune are the handoff constant k_H, the power system mass fraction β, and the handoff time t_H. The variables represent only 2 degrees of freedom, since clearly once we set β and k_H, t_H is already determined.

Under power limited scenarios, while v_T can still be tuned, it does not affect the throughput, since it only alters the behavior of the climber beyond the handoff point.

Optimization

To find the maximum throughput:

The relative payload is $m_{payload} = (1-k_H)(1-\beta)$

And the throughput is $PMT = (1-k_H)(1-\beta)/t_H$

We can now link β and k_H through t_H

Speed-limited:

Initial velocity:	$v_e = \beta \cdot \rho_P/g$
Terminal point:	$r_T = r_e(v_T/v_e)^{0.5} = r_e \cdot Q_T$
Terminal altitude:	$a_T = r_T - r_e$
Time to terminal point:	$t_T = (r_e/r_T)(r_T-r_e)/v_e = (r_e/v_e)(Q_T-1)/Q_T$
Handoff point:	$r_H = r_e/k_H^{0.5}$
Handoff altitude:	$a_H = r_H - r_e$
Time to handoff point:	$t_H = t_T + (r_H-r_T)/v_T = (r_e/v_e)(Q_T-1)/Q_T + (r_e k_H^{-0.5} - r_e Q_T)/v_T =$
	$(r_e/v_T) \cdot [Q_T(Q_T-1) - Q_T + k_H^{-0.5}] = (r_e/v_T)[Q_T(Q_T-2) + k_H^{-0.5}]$
Extracting k_H:	$k_H = [(t_H v_T/r_e) - Q_T(Q_T-2)]^{-2}$

Power-limited:

Initial velocity:	$v_e = \beta \cdot \rho_P/g$
Handoff point:	$r_H = r_e(v_H/v_e)^{0.5} = r_e \cdot Q_H = r_e \cdot (k_H)^{-0.5}$
Handoff altitude:	$a_H = r_H - r_e$
Velocity at handoff point:	$v_H = v_e Q_H^2$

Time to handoff point: $t_H = (r_e/r_H)(r_H-r_e)/v_e = (r_e/v_e)(Q_H-1)/Q_H = (r_e/v_H)Q_H(Q_H-1)$

Extracting β: $\beta = g \cdot v_e/\rho_P = (g/\rho_P) \cdot (r_e/t_H) \cdot (Q_H-1)/Q_H$

Defining: $c_H = (g/\rho_P) \cdot (r_e/t_H)$

Solving for k_H: $k_H = Q_H^{-2} = [c_H / (c_H-\beta)]^{-2}$

It is possible to express $dP/d\beta$ is closed form, but the resulting expression is only solvable numerically. For the same effort, it is more interesting to directly numerically optimize $PMT(\beta)$.

Results

The worksheet is implemented in MS Excel, and works with the built-in numerical solver to yield optimal values for PMT in respect to β. Below is one instance of the optimization, for ρ_P = 1500 and v_T = 80. The source worksheet is available online at http://www.ISEC.org/CLIMB/Vol2No1/SEPSAO.xls.

Earth radius	r_e	m	6.4E6	6.4E6	6.4E6	6.4E6	6.4E6	6.4E6	r_e		
Earth rotation	ω_e	rad/sec	7.3E-5	7.3E-5	7.3E-5	7.3E-5	7.3E-5	7.3E-5	ω_e		
Earth gravity	g	m/sec2	9.8E0	9.8E0	9.8E0	9.8E0	9.8E0	9.8E0	g		
power density	ρ_P	Watt/kg	1500	1500	1500	1500	1500	1500	ρ_P		
terminal velocity	v_T	m/s	80	80	80	80	80	80	v_T		
delta-β		0.02	-0.04	-0.02	0	0.02	0.04				
power mass ratio	β		0.199	0.219	0.239	0.259	0.279	0.239	β	[1]	
handoff time	t_H	sec	86400	86400	86400	86400	86400	86400	t_H		
Initial Velocity	v_e	m/s	30.5	33.6	36.6	39.7	42.7	36.6	v_e	[5]	
sqrt(v_T/v_e)	Q_T		1.6	1.5	1.5	1.4	1.4	1.5	Q		
terminal point	r_T	m	1.0E7	9.9E6	9.5E6	9.1E6	8.8E6	9.5E6	r_T		
terminal altitude	a_T	km	3966	3482	3060	2688	2356	3060	a_T	[7]	
terminal time	t_T	sec	8.0E4	6.7E4	5.7E4	4.8E4	4.0E4	5.7E4	t_T		
		hr	22.3	18.7	15.7	13.3	11.2	15.7			
handoff constant	k_H		0.348	0.314	0.292	0.276	0.264	0.292	k_H	[4]	
handoff point	r_H	m	1.1E7	1.1E7	1.2E7	1.2E7	1.2E7	1.2E7	r_H		
handoff altitude	a_H	km	4454	5018	5449	5783	6045	5449	a_H	[6]	
handoff time (chk)	t_H	sec	86400	86400	86400	86400	86400	86400	t_H		
		hr	24	24	24	24	24	24			
$m_{payload}$	m_p	m_{max}	0.52	0.54	0.54	0.54	0.53	0.539	m_p	[3]	
Throughput	PMT	STU	191	195	197	196	194	197	PMT	[2]	
			1	1	1	1	1	1			

Ignore columns with red indicators

			0	0	0	0	0	0			
	C_H		0.484	0.484	0.484	0.484	0.484	0.484			
	Q_H		1.700	1.828	1.978	2.154	2.364	1.978			
handoff constant	k_H		0.346	0.299	0.256	0.216	0.179	0.256	k_H		
handoff point	r_H	m	1.1E7	1.2E7	1.3E7	1.4E7	1.5E7	1.3E7	r_H		
handoff altitude	a_H	km	4478	5300	6256	7383	8729	6256	a_H		
handoff time (chk)	t_H	sec	86400	86400	86400	86400	86400	86400	t_H		
		hr	24	24	24	24	24	24			
$m_{payload}$	m_p	m_{max}	0.524	0.547	0.566	0.581	0.592	0.566	m_p		
Throughput	PMT	STU	191	200	207	212	216	207	PMT		

Throughput	PMT	STU	191	195	197	196	194	197	PMT	[2]	

Instructions:
Enter values in the 3 parameter cells marked:
Experiment with the value in the β cell:
Control the 5 test case columns using delta-β:

The Red/Green indicators show which scenario (power or speed limited) is applicable.
(Red-Red means that a different condition (such as $v_e > v_T$) is violated)

When ready, hit alt-T,V (tools-->solver) and optimize PMT with respect to β.

The parameter space is 3-dimensional, and we are interested in multiple resultant quantities. The approach taken for aggregating the data is to hold t_H constant and plot one table per observed quantity, then experiment with other values of t_H.

This process requires a considerable amount of manual work, but gives the experimenter a good insight into the behavior of the system.

Below are the results for daily-cycle operations (t_H=86400). Power limited scenarios are shaded. Our focus is on the payload throughput (PMT). m_P and k_H are shown for "situational awareness".

[1] β

	500	700	1000	1500	2500	3500
30	0.374					
40	0.420					
60	0.450	0.381	0.295	0.220	0.150	
80	0.450	0.415	0.321	0.239	0.165	
100	0.450	0.426	0.340	0.253	0.174	0.136
120	0.450	0.426	0.355	0.263	0.180	0.141

[2] PMT

	500	700	1000	1500	2500	3500
30	101					
40	105					
60	105	135	162	186	209	
80	105	137	168	197	224	
100	105	137	172	203	233	248
120	105	137	173	208	240	256

[3] m_P

	500	700	1000	1500	2500	3500
30	0.278					
40	0.287					
60	0.288	0.369	0.443	0.509	0.571	
80	0.288	0.374	0.461	0.539	0.612	
100	0.288	0.375	0.470	0.557	0.639	0.679
120	0.288	0.375	0.475	0.569	0.657	0.700

[4] k_H

	500	700	1000	1500	2500	3500
30	0.557					
40	0.506					
60	0.477	0.403	0.372	0.347	0.328	
80	0.477	0.360	0.322	0.292	0.267	
100	0.477	0.347	0.288	0.255	0.227	0.214
120	0.477	0.347	0.264	0.229	0.199	0.185

The first observation is that once the system becomes power-limited, v_T (as expected) no longer influences the result. The second observation is that even before the system becomes power-limited, the performance only advances slowly. If we stay in the "reasonable" v_T range of 60-120 m/s, the throughput values are mostly a function of the power density.

As estimated before (using the $r_H=2r_e$ point), the weaker power systems run with $m_P \approx 0.25$, but we find out that the stronger ones reach much higher, into $m_P \approx 0.6$. Since even with $m_P = 0.6$ we only have PMT = 219, there's motivation to examine what can be gained by increasing the launch rates.

Looking at bi-daily operations ($t_H=43200$) and remembering that for the same PMT m_P will be half its previous value, we get:

[1] β	500	700	1000	1500	2500	3500	[2] PMT	500	700	1000	1500	2500	3500
30	0.440						30	114					
40	0.477	0.400					40	115	151				
60	0.477	0.466					60	115	155				
80	0.477	0.466	0.420				80	115	155	209			
100	0.477	0.466	0.450	0.342	0.235	0.183	100	115	155	210	275	338	369
120	0.477	0.466	0.450	0.363	0.251	0.196	120	115	155	210	281	352	387

The shaded region is larger since the smaller climbers need to get out of the way faster and so carry larger power systems, thus maxing out sooner. For this reason, while the higher performing systems gain up to 50% in throughput, the lower performing systems gain only about 10%. This is to be expected, since there's little point expediting the launch rate if the system is not capable of getting the climbers far enough out of the way within half a day.

Conclusions

We can draw the following table, to be used as a rough guide for the throughput available from a power system: (Throughput again is in units of m_{max}/yr, or STU)

ρ_P:	Watt/kg	500	700	1000	1500	2500	3500
Daily	STU	100	135	170	200	230	250
Bi-Daily	STU	115	155	210	275	340	370

If we assume a power system that is at least partially solar based, day-cycle operations are a solid assumptions, yielding throughputs in the range of 100 – 250 STU for power systems with a power density of 0.5 – 3.5 Watt/kg. These numbers can then be plugged into the Space Elevator Feasibility Condition and compared with acceptable characteristic time constant.

In summary, the universe has once again conspired to make a terrestrial Space Elevator feasible, but just barely so.

References

[1] Artsutanov, Y., "Into the Cosmos by Electric Rocket", *Komsomolskaya Pravda,* 31 July 1960. (The contents are described in English by Lvov in *Science,* 158, 946-947, 1967.)

[2] Artsutanov, Y., "Into the Cosmos without Rockets", *Znanije-Sila* 7, 25, 1969.

[3] Pearson, J., "The Orbital Tower: A Spacecraft Launcher Using the Earth's Rotational Energy", *Acta Astronautica* 2, 785-799, 1975.

[4] Edwards, B. C., and Westling, E. A., "The Space Elevator: A Revolutionary Earth-to-Space Transportation System", ISBN 0972604502, published by the authors, January 2003

[5] B. Shelef, "Space Elevator Feasibility Condition", *C^LIMB vol 1*, 2011

[6] B. Shelef, "Space Elevator Calculation Scrapbook", *The Spaceward Foundation*, 2008

ADDITIONAL READING

THE REAL HISTORY OF THE SPACE ELEVATOR

In October of 2006, the International Astronautical Federation (IAC) held a conference in Valencia, Spain. At that conference, Jerome Pearson presented the following paper entitled *The Real History of the Space Elevator*, a fascinating look at how this idea came about.

Mr. Pearson has graciously allowed us to reprint that paper in this Volume of C**L**IMB and for that we are very grateful.

IAC-06-D4.3.01

THE REAL HISTORY OF THE SPACE ELEVATOR

Jerome Pearson

STAR, Inc., Mount Pleasant, SC, USA, jp@star-tech-inc.com

Abstract: The space elevator was invented twice, independently, and many ideas about space elevators were developed by multiple authors and inventors, some without knowledge of the others. The history of the field is somewhat muddled, so much that some fairly recent papers have re-invented the concept. Some sources attribute the invention of the space elevator to Sir Arthur Clarke, in a novel in 1978. Others attribute it to Tsiolkovsky, writing in 1895. Part of the problem is that the original inventor of the space elevator, Yuri Artsutanov, published only in the youth-oriented *Komsomolskaya Pravda*. Even after John Isaacs and his colleagues came close to the concept of the space elevator in an article published in *Science* in 1966, there was no notice in the spaceflight engineering community. British authors Collar and Flower almost re-invented the concept in the *JBIS* in 1969, but again the spaceflight engineering community did not pick up the idea. The independent invention by Jerome Pearson, published in *Acta Astronautica* in 1975, finally made the concept widely known. Now that there are significant efforts underway to develop the materials required to actually build a real space elevator, it is worthwhile to review the convoluted and somewhat mysterious history of the space elevator concept.

I. INTRODUCTION

Humans have had yearnings to escape the bonds of Earth and reach for the stars since time immemorial. The pyramids of Egypt were presaged by ziggurats in Mesopotamia that led to the story of the Tower of Babel in the Bible. But it has only been in the last century or so that such yearnings could be put into scientific or engineering terms.

In 1895, the Russian space pioneer Konstantin Tsiolkovski published two pamphlets[1] called "Dreams between Earth and Sky" and "On Phobos." In these scientific speculations couched as science fiction, Tsiolkovski examined the problems of traveling into space, and conceived tall towers and globe-circling trains to replace rockets.

Using the Eiffel Tower as a model, Tsiolkovski imagined towers reaching into space, and discovered the balance point at which gravity seems to disappear, which is the synchronous altitude. However, in these pamphlets, Tsiolkovski was performing "thought experiments," and felt that building real towers into space was impossible[2].

Another great Russian space pioneer was Friedrich Tsander, who designed the first Russian liquid-fuel rocket. Tsander is said to have conceived of the lunar space elevator[3] in the 1920's, but an English translation of his writings may not exist.

II. CONCEPTS AND ORIGINATORS

Apparently the first person to think of building a space elevator as an engineering project was Yuri Artsutanov of Leningrad (now St. Petersburg), in 1960[4]. In 1959, as a student at the St. Pete school, he was given a sample of a very strong material by a colleague. He immediately thought that such a strong material could be suspended over a very long length, and being interested in spaceflight and science fiction, he thought of using it to make a "cosmic railway," which would reach to geosynchronous orbit[5].

Artsutanov was still a student, and did not try to publish his idea, but a journalist friend wrote an article in the youth supplement to the Sunday *Komsomolskaya Pravda*, and many of his ideas appeared in the short article. The concept is shown in **Figure 1**.

Figure 1. Artsutanov Space Elevator

The space elevator concept did not catch on in the West, however, and there was no notice taken until the near-invention by Isaacs et al. (see below) published in *Science* magazine in 1966. Even then, there was no notice taken by the spaceflight community, although several university professors used it as a physics problem for their students.

Even such a spaceflight enthusiast as Sir Arthur Clarke did not realize the significance of the Artsutanov concept. The idea was depicted in a volume of paintings[6], and Clarke saw the picture, but did not realize its significance.

The other invention of the space elevator was by the author of this paper (**Figure 2**) in 1975[7]. The origin of this idea was from Arthur Clarke's testimony before Congress in 1969, describing geostationary satellites as "perched above imaginary towers 22,000 miles high." When I heard this, I thought: why not drop a real line down to the ground, and make a space elevator? After completing the concept and realizing that an exponential taper was required, I tried unsuccessfully for 5 years to publish before the courageous editor A. K. Oppenheim published it in *Acta Astronautica*.

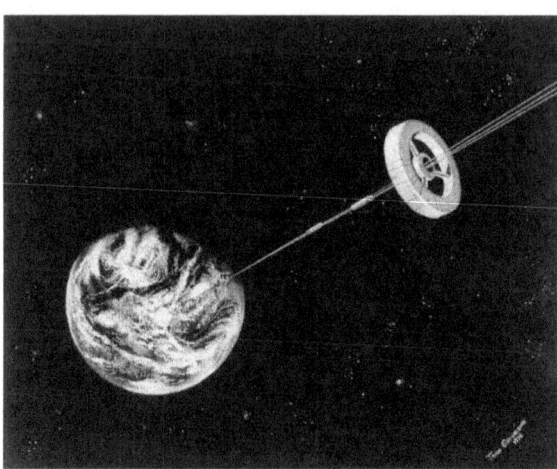

Figure 2. Pearson Space Elevator

Once the Orbital Tower paper was published by *Acta Astronautica*, there was great interest in the spaceflight community, and importantly, by Arthur Clarke (**Figure 3**). He wrote a novel based on the idea, and the two of us corresponded many times until the book came out in serial form in 1978 and in book form in 1979[8].

Figure 3. Sir Arthur Clarke and Jerome Pearson, Sri Lanka, 1988

In August of this year, the author had the distinct pleasure of traveling to St. Petersburg, Russia, to visit with Yuri Artsutanov and to discuss the origins of our space elevator concepts. We met outside the Hermitage Museum, as shown in **Figure 4**.

Figure 4. Yuri Artsutanov and Jerome Pearson, St. Petersburg, 2006

III. <u>NEAR INVENTORS</u>

In the 1960's there were several near inventions of the space elevator. The first and closest was by a group of oceanographers at the Scripps Institution of Oceanography at La Jolla, California, led by John Isaacs[9]. They were experienced in the use of long, tapered cables in the open ocean, and applied it to space. Due to either editorial pressure or timidity, they did not propose a practical structure, but a microscopic pair of diamond filaments that would be severed by micrometeoroids in a matter of moments. They also proposed that payloads be lifted into space by "jockeying" the filaments up and down, which cannot be scaled up to carry large payloads. In spite of this conservatism, the editor prefaced the article with the unique caveat that he didn't think it would work!

After Syncom 1, the first geostationary communication satellite, was launched in 1962, there were concerns that the great distance to GEO required too much power, and too much signal delay time. In response to these problems, two separate groups proposed lowering satellites from GEO on long, tapered cables to reach lower altitudes. The first was by Sutton[10] in the US, in 1967. Sutton proposed a cable 10,000 miles long, lowering the satellite to half of GEO altitude, but did not logically extend it to the Earth's surface to make a space elevator.

The second, by Collar and Flower[11], was even more interesting. They also proposed hanging satellites below GEO on tapered cables, and even noted that such a cable could support a small payload down to the Earth's surface. But they stopped there, not realizing that this was a space elevator.

IV. <u>RELATED CONCEPTS</u>

In addition to the near-inventions of the space elevator, there are several related ideas that are generalizations or modifications of the concept. The classical space elevator is a satellite that rotates in the same time that it orbits the planet, and touches down at one point. If the satellite were lower and rotated faster, its arms could touch down on the equator at two, three, or more points, like the spokes of a giant wheel rolling around the Earth. Artsutanov[12] originated this concept in a 1969 paper, but like his earlier concept, it was not known in the West.

Working independently, Hans Moravec[13] at Stanford University, inspired by Professor John McCarthy, came up with the same idea, and found the optimum length for minimum mass. He also proposed freely rotating "velocity banks" in orbit to catch and throw payloads, providing delta-V for orbit transfers, like planetary flybys for gravity assists.

In the 1980's several new ideas were developed. In addition to static space elevators, there can also be dynamic space elevators that depend on the inertia of moving masses for support. Paul Birch[14] in England proposed shorter space elevators suspended from an orbiting ring in LEO. This hollow cylinder, like a giant donut, would be supported by a high-speed wire inside, moving at much higher than orbital velocity, and providing an upward electrodynamic force on the cylinder, from which the short space elevators would be suspended.

A variation on this approach was proposed by Benoit Lebon[15] in France, in the form of an orbiting ring of particles. Loops about the ring could support and accelerate spacecraft electrodynamically, although there are questions about the stability of such a grain stream ring.

A similar idea is the "Launch Loop" proposed by Keith Lofstrom[16]. This is again a hollow tube supported by a high-speed wire inside, like the Birch orbital ring, but it is shorter and anchored on the ground at both ends. The high-speed wire completes its loop on the ground, and the upper part is levitated by the moving wire.

Perhaps the most audacious of the dynamic space elevators is the "Star Bridge" proposed by Rod Hyde[17] and Lowell Wood. This is a purely vertical tube, with a device at the ground that launches high-speed magnetic rings upward, where they are turned and accelerated downward by a similar device at the top. The reaction force supports the tower, and it can theoretically be built to any height.

All of these concepts would allow the space elevator to be anchored off the equator. This can also be done with the classical space elevator, by swinging or pulling the lower end away from the equator and then anchoring it. This was proposed by Pearson[18] in 1976, and again by Gassend[19] in 2004.

If the space elevator is anchored off the equator, there is an additional stress on the ribbon, and a maximum latitude that can be reached, depending on the gravity field and the strength of the material. Bolonkin[20] suggested a tower at the pole, just tall enough that the hanging ribbon does not contact the ground. This is impractical for the Earth, and probably for the Moon also, but it might be workable for small asteroids.

Finally, the lunar space elevator, balanced about the Lagrangian points L1 or L2, was first proposed by Pearson[21] in 1977 and independently by Artsutanov[22] in 1979. Because the Moon's mass is so much less than Earth's, the lunar space elevator could be constructed of existing materials. The lunar space elevator could be quite useful in the development of lunar resources, as shown by Pearson et al[23].

V. KEY PROBLEMS

If we are to see the realization of the Earth space elevator, there are several very important problems that must be solved. The first, of course, is the very high material strength, which is being addressed by the investigation of composites made of carbon nanotubes. Even if the materials problem is solved, however, there remain other very serious problems.

Because the space elevator is fixed with respect to the Earth, and extends all the way from the equator to 100,000 km, the orbits of all other Earth satellites (and debris) will sooner or later intersect with the elevator. The resulting high-velocity collisions will destroy the colliding satellites, which may have cost billions of dollars to build and launch. The collisions could also damage or destroy the space elevator as well.

Two things must be accomplished to avoid these disasters. First, Earth orbit must be cleaned up of all debris larger than a few centimeters. This is the center of attention of international bodies, including the International Academy of Astronautics, and there are potential techniques to accomplish it, including an electrodynamic tether proposed by this author[24].

Second, we must devise ways to reliably avoid collisions between the space elevator and high-value Earth satellites, including classified satellites whose positions are not published. This will be incumbent on the space elevator itself, and is a most serious problem. Brad Edwards[25] has suggested that we induce lateral oscillations in the space elevator ribbon, timed precisely to avoid other satellites, but this technique has never been demonstrated, and there may be serious questions about its efficacy.

This leads to the third problem, and that is the perceived difficulty of building the space elevator, because of its gigantic scale. The basic idea is so audacious that in the early days it was difficult to convince editors to even publish papers on the space elevator. Isaacs only published with a heavy caveat from the editor, and the orbital tower paper took 5 yrs to find a willing editor.

Further, the history of space tethers, which are basically just short versions of the space elevator, has not been encouraging. There have some high-visibility failures, such as the NASA Tethered Satellite System Space Shuttle launches, TSS-1 and TSS-1R. The TSS-1 tether failed to deploy, and it was re-flown. On TSS-1R, the electrodynamic tether arced and was severed, losing the satellite. Since then, it has been extremely difficult to convince program managers to trust space tethers for any important mission.

VI. **POTENTIAL SOLUTIONS**

To succeed in building and launching the space elevator, it will be necessary to overcome each of these problems. An incremental approach seems to be required.

First, we need to have a series of successful space tether missions that will build confidence in their safety, utility, and controllability, as I suggest in a companion paper at this symposium[26]. Ideally, these precursor tether missions should demonstrate high enough payoffs that space tethers are accepted as part of the operational spacecraft mix. One way to do this is to use space tethers to remove the very debris that threatens the space elevator, as my colleagues and I have proposed for LEO debris[22], and as Chobotov[27] has proposed for GEO debris. Another is to use space tethers to support the new NASA Exploration Systems in returning to the Moon and going on to Mars. One way is to use rotating space tethers to provide artificial gravity for life sciences studies and for long Mars missions[28].

Finally, we must address the problem of the controllability and benign dynamic behavior of space elevators. This may require that we first emplace much smaller versions of the space elevator on the

Moon or on near-Earth objects. In the process of retrieving lunar and asteroid resources, we will also be able to demonstrate the control of a 100,000-km-long space structure. This can be on a much smaller and cheaper scale than the Earth space elevator, so we will not be risking the loss of the entire program.

Once we have gained the experience with these multiple space tether missions, and demonstrated their safety and payoffs all the way through asteroidal and lunar space elevators, the world will be ready to accept the practicality of the Earth space elevator.

VII. CONCLUSIONS

For thousands of years, there were ancient dreams of towers to heaven. Between 1895 and 1930, Konstantin Tsiolkovski and Friedrich Tsander put such speculations on a scientific basis. Finally, the space elevator as we currently envision it was invented by Yuri Artsutanov in 1959, and independently by Jerome Pearson in 1970.

Confusion about the real history of the space elevator was compounded by several near inventions in the 1960's, and by science fiction and journalistic interest that focused on Sir Arthur Clarke and Konstantin Tsiolkovski. In addition, there are still new papers being written that re-invent the space elevator without being aware of the now-widespread information available.

The purpose of this paper has been to shed light on this murky history, and also to point the way to actually building the space elevator.

Acknowledgments

The author is indebted to most of the principals in the space elevator saga and others for personal communications and information, including Yuri Artsutanov, Arthur Clarke, Hans Moravec, John Flower, colleagues of the late John Isaacs, Eugene Levin, and others.

REFERENCES

[1] Tsiolkovski, Konstantin E., Grezy O Zemle I Nebe [i] Na Veste (Speculations of Earth and Sky, and On Vesta, science fiction works, 1895). Moscow, Izd-vo AN SSR reprint, 1959.

[2] Pearson, J., "Konstantin Tsiolkovski and the Origin of the Space Elevator," IAF-97-IAA.2.1.09, 48th International Astronautical Congress, Turin, Italy, 6-10 October 1997.

[3] Levin, Eugene, personal communication, 2006.

[4] Artsutanov, Y., "Into the Cosmos by Electric Rocket," Komsomolskaya Pravda, 31 July 1960.

[5] Artsutanov, Yuri N., Personal letter to Arthur Clarke, April 1979.

[6] Leonov, Alexei, and Sokolov, Andrei, Live Among the Stars, Moscow, 1967.

[7] Pearson, J., "The Orbital Tower: A Spacecraft Launcher Using the Earth's Rotational Energy," Acta Astronautica Vol. 2, pp. 785-799, 1975.

[8] Clarke, Arthur C., The Fountains of Paradise, Harcourt Brace Jovanovich, New York, 1979.

[9] Isaacs, J., Vine, A. C., Bradner, H., and Bacchus, G. E., "Satellite Elongation into a True Skyhook," Science, Vol. 151, pp. 682-3, 1966.

[10] Sutton, G. W., and Diederich, F. W., "Synchronous Rotation of a Satellite at less than Synchronous Altitude," AIAA Journal, Vol. 5, No. 4, pp. 813-815, April 1967.

[11] Collar, A. R., and Flower, J. W., "A (Relatively) Low Altitude Geostationary Satellite," Journal of the British Interplanetary Society, Vol. 22, pp. 442-457, 1969.

[12] Artsutanov, Y., "Into the Cosmos without Rockets," Znanije-Sila 7, 25, 1969.

[13] Moravec, H., "Non-Synchronous Orbital Skyhook," Journal of the Astronautical Sciences, Vol. 25, No. 4, pp.307-322,1977.

[14] Birch, P. M. "Orbital Ring Systems and Jacob's Ladders," JBIS, Vol. 35, No. 11, pp. 475-497, November 1982.

[15] Lebon, B. A., "Magnetic Propulsion along an Orbiting Grain Stream," Journal of Spacecraft Vol. 23, pp. 141-143, 1986.

[16] Lofstrom, K., "The Launch Loop—A low-cost Earth-to-orbit Launch System," 21st Joint Propulsion Conference, AIAA Paper 85-1368, Monterey, July 8-10 1985.

[17] Hyde, R. A., "Earthbreak: Earth to Space Transportation," Defense Science 2003+, Vol. 4, pp. 78-92, 1985.

[18] Pearson, J., Letter to Arthur Clarke, 6 November 1976.

[19] Gassend, B., "Non-Equatorial Uniform-Stress Space Elevators," 3rd Annual Space Elevator Conference, Washington, DC, June 2004.

[20] Bolonkin, A. A., "Non-Rocket Earth-Moon Transport System," COSPAR 02-A-02226, 34th Scientific Assembly of the Committee on Space Research (COSPAR), World Space Congress, Houston, Texas, 10-19 October 2002.

[21] Pearson, J., "Anchored Lunar Satellites for Cislunar Transportation and Communication," Journal of the Astronautical Sciences, Vol. 27, No. 1, pp. 39-62, 1979.

[22] Artsutanov, Yuri, "Trakka 'Luna¬-Zemlya'," ("Railway 'Moon¬-Earth'"), Technika Molodishi, No.4, p. 21, 1979.

[23] Pearson, J., Levin, E., Oldson, J., and Wykes, H., "The Lunar Space Elevator," IAC-04-IAA.3.8.3.07, 55th International Astronautical Congress, Vancouver, Canada, October 2004.

[24] Pearson, J., Carroll, J., Levin, E., Oldson, J., and Hausgen, P., "Overview of the Electrodynamic Delivery Express (EDDE)," AIAA 2003-4790, 39th Joint Propulsion Conference, Huntsville, 20-23 July 2003.

[25] Edwards, B. C., and Westling, E. A., The Space Elevator: A Revolutionary Earth-to-Space Transportation System, Published by the authors, 2002.

[26] Pearson, J., and Oldson, J. C., "High-Payoff Space Tethers," IAC-06-D4.3.06, 57th International Astronautical Congress, Valencia, Spain, 2-6 October 2006.

[27] Chobotov, V. A., "Disposal of Geostationary Satellites by Earth-Oriented Tethers," IAC-04-IAA.3.8.2.05, 55th International Astronautical Congress, Vancouver, Canada, October 2004.

[28] Pearson, J., MARCUS briefing to NASA Office of Biological and Physical Research, April 2004.

SPACE ELEVATOR INITIAL CONSTRUCTION MISSION OVERVIEW

David Lang is an American Engineer and Consultant who, among his many activities, has written extensively about the Space Elevator and Tethers in space. He is the creator of GTOSS, the Generalized Tether Object Simulation System, a computer language that allows the user to simulate the actions of Tethers.

As part of his GTOSS Space Elevator simulations, he has written several papers, one of which is reproduced here with Mr. Lang's gracious permission.

You can find all of these papers and more, including several simulations at his website: http://home.comcast.net/~GTOSS/GTOSS_and_Space_Elev.html.

You can learn more about David Lang and GTOSS at http://home.comcast.net/~GTOSS/.

Space Elevator Initial Construction Mission Overview

David D. Lang[1]

[1]*David D. Lang Associates, Seattle WA.*

Abstract: This paper presents an overview of a proposed GEO-originating deployment mission flight scenario currently under consideration to accomplish the initial construction of a space elevator. Much as a suspension bridge's initial strand of cable must be established, so must the elevator's first strand of vertical ribbon and ballast mass be erected. Results of dynamic simulations of initial deployment accomplished using the Generalized Tethered Object Simulation System (GTOSS) software tool are presented via discussions and summary graphs. A brief overview of dynamic models that constitute GTOSS is presented, and the physical configuration of the elevator as it manifests itself within GTOSS is characterized. A general discussion of orbital dynamics challenges facing this initial deployment process is presented, with emphasis on dynamic control issues and implications on the two space craft delegated to performing the deployment. Finally, a proposed control strategy is presented and simulated to demonstrate the possibility of a GEO originating deployment.

1. Introduction

Currently two different approaches to deploying the initial elevator ribbon are identified. Both start with a space craft containing (either initially or via build-up by multiple courier-missions) the Ribbon, Ballast mass, Ballast-end controller (GEO craft), ribbon Anchor-end controller (Deploy craft), and propulsion-control systems. While they differ in starting point and maneuver strategy, they both must face the dynamics challenges of extreme tether extension. The two scenarios are:

(a) Start with a space-craft at **GEO**, thus deploying Ribbon downward from there, in conjunction with a coordinated *upward* maneuvering of the GEO craft). See Reference 1.

(b) Start with a space-craft in **LEO**, deploying the Ribbon and Ballast mass upward, creating a *system* with ever-longer orbiting period, until the configuration grows to include GEO altitude and beyond, and manifests an "orbital period" corresponding to earth rotation rate. See Reference 6.

This paper specifically explores the dynamics of the GEO deployment mission. A proposed deployment control strategy is presented that serves to expose the nature of the dynamics challenges inherent in this mission, and explores some of the intrinsic ingredients that might constitute a successful deployment mission design.

2. GTOSS Overview

The **G**eneralized **T**ethered **O**bject **S**imulation **S**ystem is a time-domain dynamics simulation code, conceived by the author in 1982 to provide NASA with a tool to simulate dynamics of combinations of space objects and tethers for flight safety certification of the Shuttle Tethered Satellite System missions. Since then, GTOSS has undergone continuous evolution and validation, being applied at some stage in the formulation of virtually every US tethered space experiment flown or proposed to date. Below is an overview of GTOSS features.

• Multiple bodies, with 3 or 6 degrees of freedom, connected in arbitrary fashion by multiple tethers, subject to natural planetary environments, including standard earth models as well as more rudimentary models for the other planets.

• Tethers represented by either *massless* or *massive* models. The *massive* tether model is a *point synthesis* approach, each tether employing a constant number of up to 500 nodes, specifiable by tether (500 is a *soft* system-configurable limit).

• All tethers can be deployed from, or retrieved into, objects by user-definable scenarios. The tether model includes momentum effects of mass entering or leaving the domain of the tether itself, and produces related forces on objects deploying and retrieving the tether material.

• Tethers can possess length-dependent non-uniform attributes describing elastic cross section, aerodynamic cross section, and lineal mass density.

• Tethers are subject to distributed forces from: Subsonic and hypersonic aerodynamics; Electrodynamics of current interaction with magnetic fields using current-flow models incorporating effects of insulated or bare-wire conductors interacting with a plasma environment model. With a ribbon-to-plasma contact model, grounding-current in a conducting elevator may possibly be assessed.

• Tethers experience thermal response, gaining heat under the influence of solar radiation, earth albedo, earth infrared radiation, aerodynamics, and electrical currents; heat loss occurs through radiative dissipation.

• Tethers can be severed at multiple locations during simulation.

• Objects and tethers can be initialized in many ways, including creating stabilized extremely long tether chains, attached to and rotating with a planet (a space elevator) with due consideration for variation in longitudinally non-uniform tether properties.

• GTOSS creates a database containing results of response to the user-defined material configuration, initialization specifications, and environmental options; this permanent data base can then be *post processed* to produce a wide variety of result displays, from tabular data, to graph plots, to animations.

3. Deployment Configuration Model and Mechanics

For the GEO deployment mission, the topology consists of a Deploy craft, (as the ribbon's lower body) that proceeds earth-ward during deployment and a GEO craft (as the ribbon's upper body) that proceeds ballast-ward; it is the GEO craft that contains the ribbon characterized by a dual tapered design. Deployment occurs initially by downward ejection of the Deploy craft attached to the earth-end of the tapered ribbon (emerging first along with the Deploy craft), leaving the ballast-end and upper portions of the tapered ribbon stowed within the GEO craft (as the GEO craft rises toward the ballast altitude). Figure 1 below depicts schematic snapshots of the system at various stages of deployment.

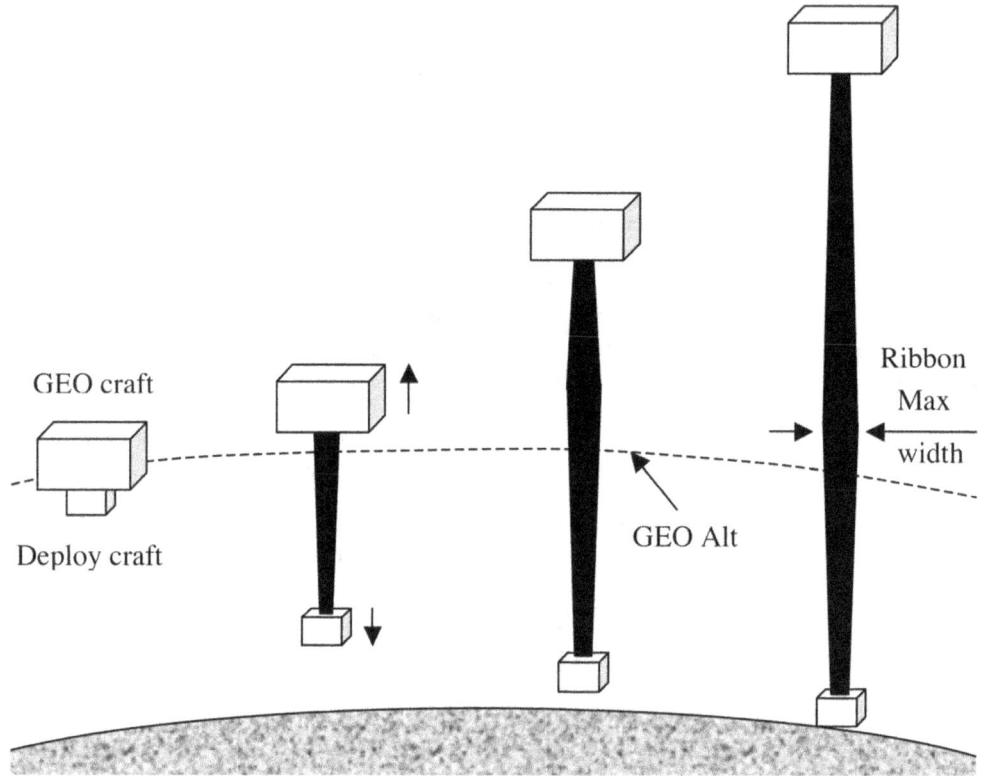

Figure 1. Snap Shots of Deployment Topology

4. Physical Properties of Initial Deployment Ribbon

GTOSS characterizes the elevator ribbon with length-varying attributes of density, elastic-area, modulus, aerodynamic-area, and damping that correspond to baseline ribbon design (described in References 1, 2, 3, and 4), but, appropriately modified to reflect a proposed initial deployment configuration. This initial deployment envisions two 20 metric ton reels of ribbon being deployed from a GEO craft (deployed simultaneously as <u>one</u> ribbon, 10 cm wide in the GTOSS simulation). By assuming that this initial ribbon would have a longitudinal taper design identical to the mature elevator, then, corresponding ribbon properties for the initial deployment mission can be derived as a scaled version of the mature ribbon. Baseline mass of the mature elevator ribbon is 825 tons, so the ratio between the initial 40 tons of ribbon and the mature ribbon yields the *lineal density ratio* of the initial ribbon in comparison to the mature ribbon. This is:

$$\frac{\rho_i}{\rho_m} = \frac{M_i}{Mm} = \frac{40}{825} = 0.0485$$

The ribbon elastic area is also scaled per this ratio. Based on the *scaled* elastic cross sectional area profile and a nominal value of the ribbon material's bulk density of 1.3 gm/cm³ (CNT), the lineal density profile was derived for use in GTOSS. Figure 2 depicts these resulting properties.

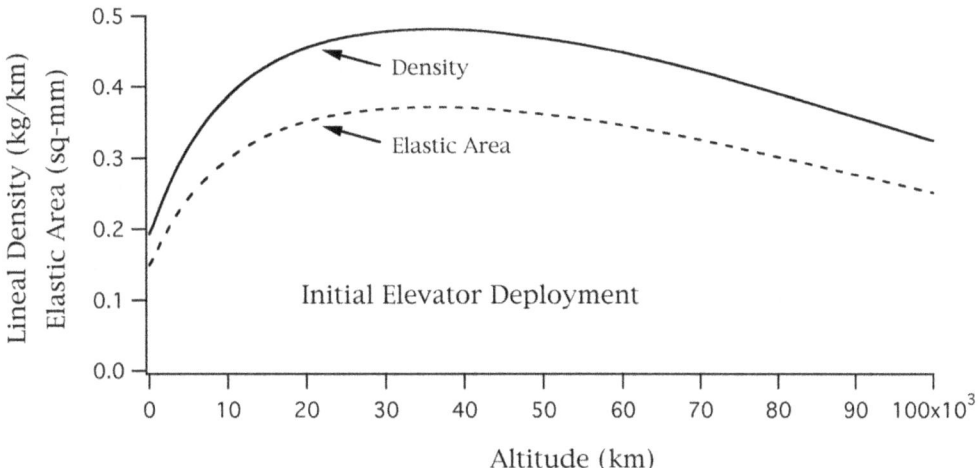

Figure 2. Ribbon Elastic Cross Sectional Area and Lineal Density vs Altitude

This above data combined with a nominal Young's modulus of 1,300 GPa (assumed for the CNT ribbon material) fully defined the ribbon's *static* elastic properties. Dynamic attributes were characterized by a material-intrinsic, strain-proportional damping factor, $\beta = 0.02$, where:

$$\sigma = E(\varepsilon + \beta \frac{d\varepsilon}{dt})$$

and,
σ = stress
ε = strain
E = Young's Modulus

5. Physical Properties of GEO Craft and Deploy Craft

Both the GEO craft and Deploy craft were simulated as 3 degree of freedom objects, thus no attitude control considerations were involved; this was deemed appropriate considering the study was only focused on the overall orbital behavior of the deploying *distributed mass* system, and the fact that the attitude dynamics of the end-bodies would essentially be uncoupled from the gross orbital dynamics. Both of these craft can loose mass due to propellant expenditure. Only the GEO craft will loose mass due to ribbon deployment. In these preliminary studies, mass loss specifically due to propulsion was inhibited because of many factors; for instance the non-optimal nature of these initial controllers *plus* the lack of propulsion technology definition that would be employed (with attendant specific impulse uncertainty) would likely produce misleading propellant usage estimates.

The initial total mass of the (upper) **GEO craft** was 69,000 kg, of which 40,000 kg is ribbon mass and 29,000 kg is ribbon deployment mechanisms, control systems, thrusters, and propellant. The initial total mass of the (lower) **Deploy craft** was 1,500 kg. This entire mass is delegated to anchor-station grappling hardware and fixtures, control systems, thrusters, and propellant.

6. Uncontrolled Natural Deployment Tendencies

It is naïve to assume that a ribbon can simply be dropped straight down from a geo-synchronous station to the earth's surface, thus effecting the initial construction of the elevator. This part of the paper addresses the natural dynamic tendencies exhibited by a GEO-positioned craft attempting a totally uncontrolled vertical deployment of ribbon. While this behavior simply reflects the response of a *greatly-extending system of connected particles* in an inverse square central force field, it nevertheless manifests tendencies indicative of what must be dealt with to successfully deploy an elevator ribbon. The GEO craft is positioned in a geo-synchronous orbit. The Deploy craft is ejected vertically down at 200 km/hr, attended by a constant ribbon deploy rate slightly in *excess* of 200 km/hr, intentionally creating a transient slack ribbon condition. No attempt is made to exercise control of either *end body*. Prior to the ribbon going taut, the Deploy craft is simply in a free Keplerian orbit moving posigrade relative to the GEO craft (that essentially remains at GEO). At about ½ day, gravity has accelerated the Deploy craft away from the GEO craft removing ribbon slack and resulting in a minor impact loading event. Following this, *range rate* between the Deploy craft and GEO craft tracks *deploy rate* until near the end of deployment, when high tension comes into play. This is shown in Figure 3 which compares the constant deploy rate to the range-rate between the GEO- and Deploy crafts.

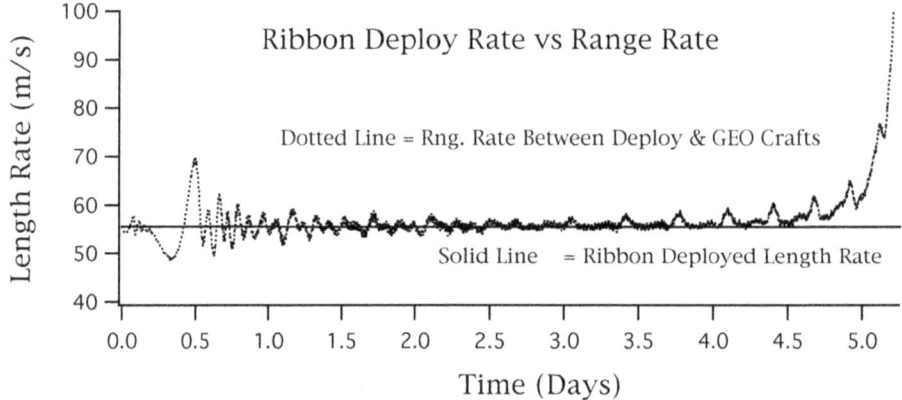

Figure 3. Ribbon Deploy Rate vs Range Rate

The steady increase in tension at both ends of the ribbon is seen in Figure 4 below. The higher tension at the GEO craft is responsible for pulling the GEO craft earthward and a resulting significant posigrade motion of the GEO craft with respect to its initial geosynchronous position. The sharp tension increase at the Deploy craft near the end of deployment is due to its diving ever more rapidly into the inverse-square gravity-well. This case clearly illustrates the potential for a deployment to end disastrously in a crash to earth!

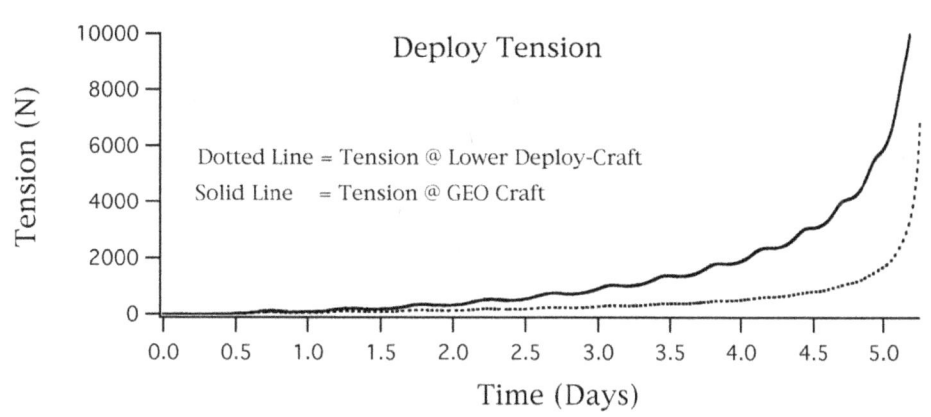

Figure 4. Tension at Upper and Lower Ends of Ribbon

The altitude state of both end-objects, shown in Figures 5 and 6, exposes the sharp increase (at 4.5 days) in the Deploy craft's accelerating encounter with gravity, *dragging everything down with it*. Consequently, the GEO craft vacates its geo-synchronous condition (loosing altitude as ribbon tension increases, pulling it earthward), and simultaneously moves through about 180 degrees of posigrade earth longitude prior to the system's eventual plunge to earth.

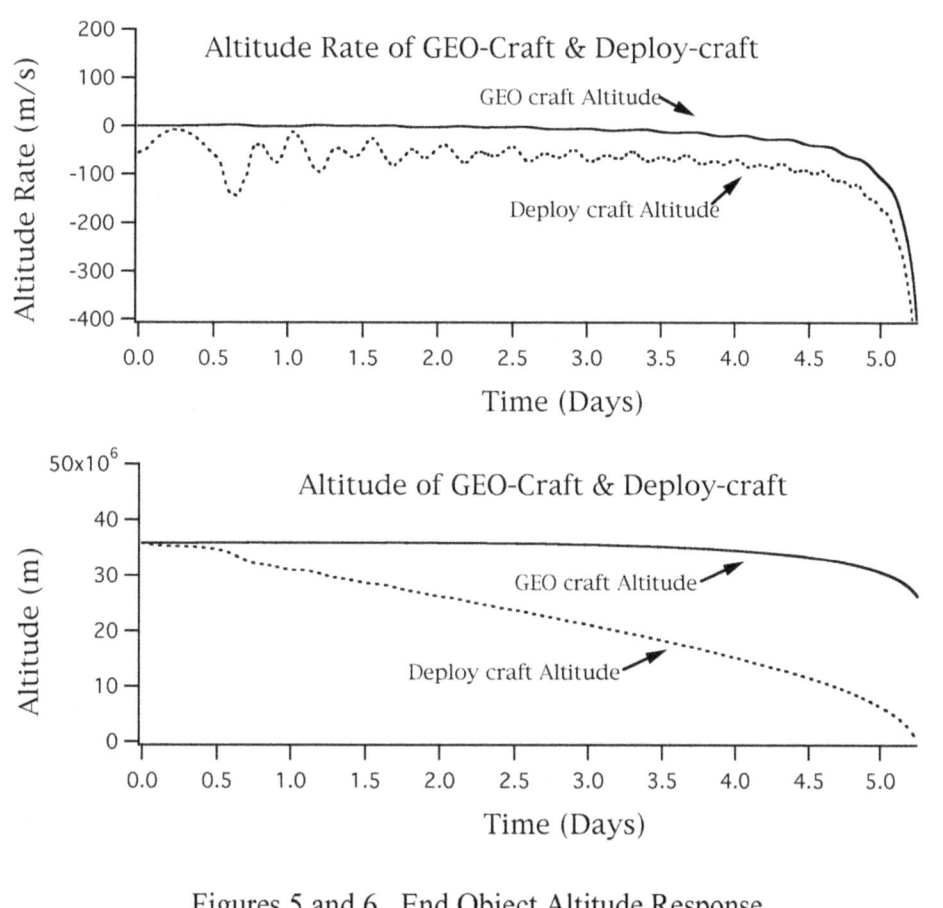

Figures 5 and 6. End Object Altitude Response

The Deploy craft dips deeply into the *gravity well* at 4.5 days, shown in Figure 7.

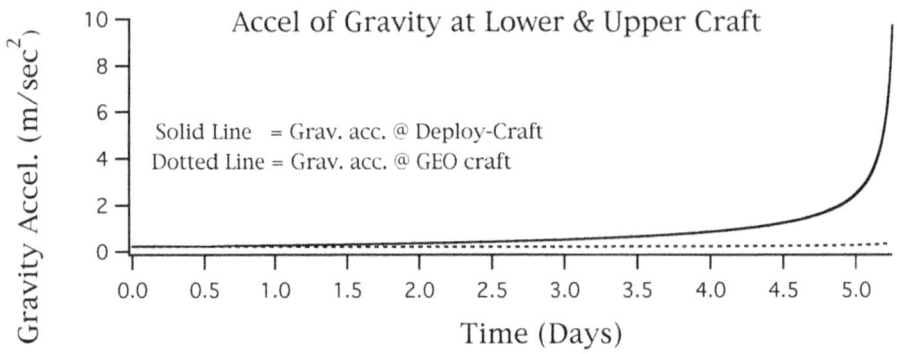

Figure 7. End Object Acceleration of Gravity

Figure 8 shows the earth longitude traversed by the GEO craft.

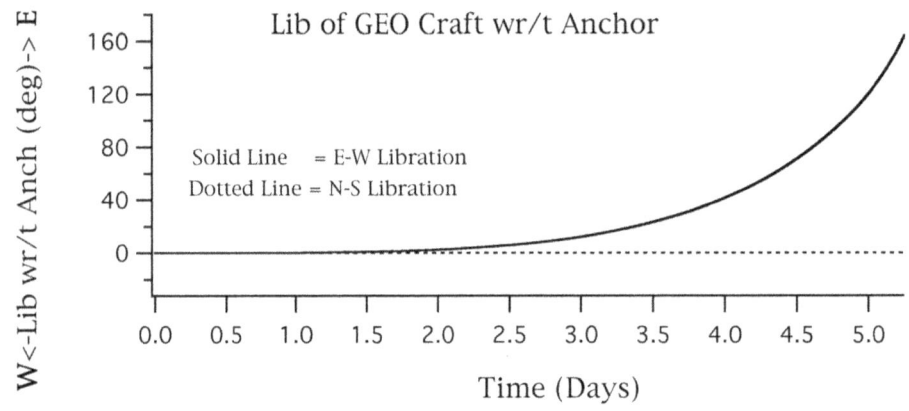

Figure 8. Earth Longitude Traversed by GEO Craft

As soon as tension manifests itself (at about ½ day) Figure 9 shows that the Deploy craft starts to librate with respect to the GEO craft, typical of tether deployment behavior. But this libration naturally tends toward zero amplitude, a feature that can be used to advantage in the mission design.

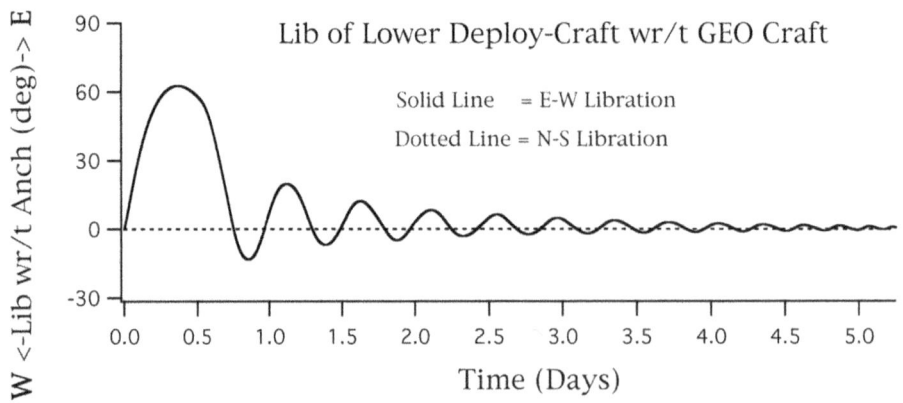

Figure 9. Libration of **Deploy** craft with-respect-to the **GEO** Craft

7. A Controlled Fly-Away From Earth

This is an example of a system under the action of a deployment controller, that while intended to equilibrate the rising vertical tension by maneuvering the GEO craft to higher altitudes, in fact over-compensates as a result of oscillatory coupling between control modes and amongst natural frequencies inherent in the elevator elastic system. The final cause of the fly-away is due to an inadequate amount of elevator mass engaging the *gravity-well* due to the instability and a subsequent ever increasing altitude rate for the elevator. These control instabilities were caused by improper gain selections and inadequate sensor filtering to insure overall stability. Figure 10 shows that the deploy rate is modulated in an attempt to control the deploy craft altitude rate as it approaches earth.

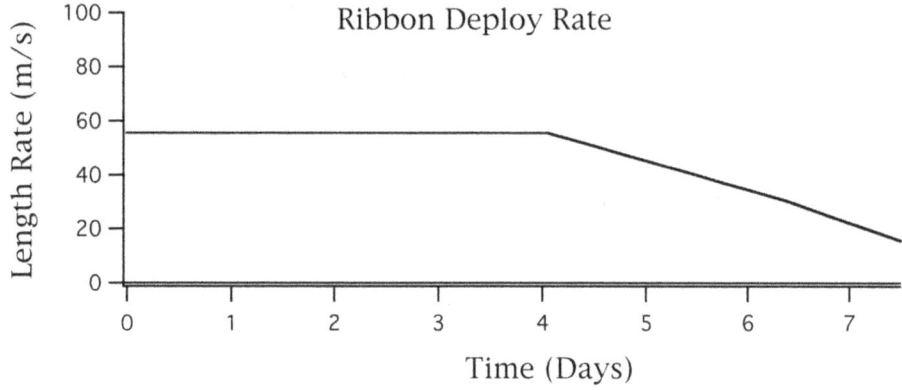

Figure 10. Deploy Rate

The ultimate failure of this deployment is clearly seen in the GEO and Deploy craft altitude histories shown in Figure 11. The GEO craft is rising, attempting to equilibrate tension, but at about 6 days into the mission, a vertical instability starts manifesting itself, after a few cycles of which, the system instability overwhelms the control effectors, and the system irretrievably departs controlled flight!

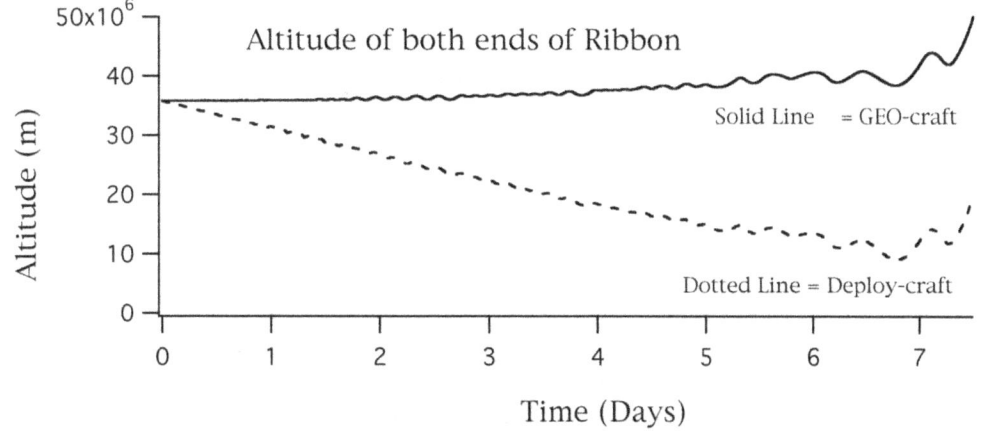

Figure 11. Altitude of GEO craft and Deploy craft

In Figure 12, it is seen that the system as a whole has failed to "bite into the gravity well" sufficiently to prevent a centrifugal departure.

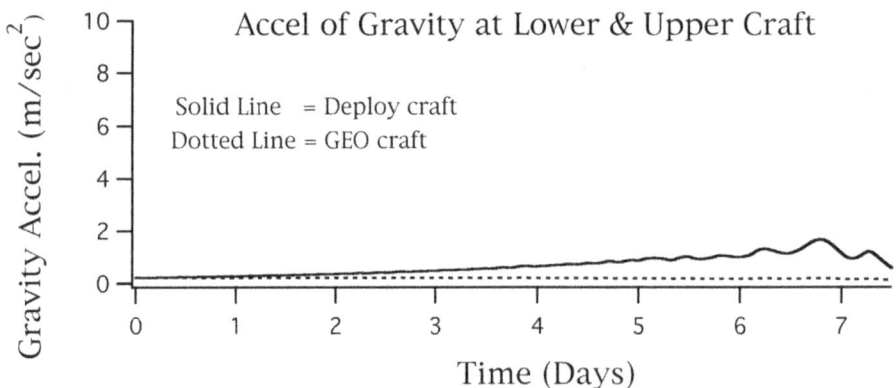

Figure 12. Acceleration of Gravity at GEO craft and Deploy craft

Tension shown in Figure 13 illustrates inappropriate control system design that is exciting longitudinal ribbon dynamics and the system. These variations are indicative of the need for an elevator ribbon deployment control system design to be able to reject undesirable frequencies in the tension signal so as to deduce *intrinsic tension level*, against which the GEO craft must fly a compensating equilibrium-altitude maneuver.

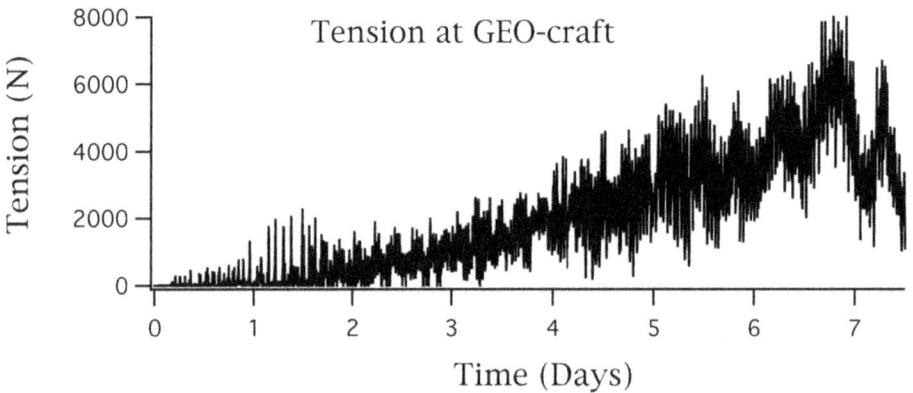

Figure 13. Tension at GEO craft

8. The Deployment Venue

The process of deploying a ribbon with physical extension on the order of the space elevator (earth to 100,000 km) is found to be a delicate control process. Little of the knowledge-base derived from actual orbital tether operations to date has bearing on this procedure due to a host of attributes that make this process unlike any yet attempted by mankind. To understand technical issues facing deployment, one must have a grasp of the physical factors inherent in this process, *the deployment venue*, as outlined below.

• Regardless of whether deployment starts at LEO or at GEO, the final *configuration* must be a vertical ribbon extending from near earth up to *centrifugally effective altitudes* at which net ribbon tension can be maintained to produce at least a condition of *neutral buoyancy*. Such a configuration (until actually attached to the earth) is neutrally stable; a perturbation that moves the (neutrally) stable-state *upward* results in a net downward force reduction that encourages the tendency. This is because every particle of mass becomes attracted less toward the earth by virtue of the *inverse radius-squared* gravity field, thus a net reduction in gravity force ensues. Countering this gravity reduction, the corresponding particles are subject to less *centrifugal effect*, however this varies as the *inverse radius*, thus restoration due to gravity is decreasing faster than the centrifugal effects are decreasing, *all of which contributes to the initial upward perturbation combining to move the system higher*. Conversely, if the ribbon moves closer to earth, just the opposite of all the above ensues, and the net effect is to pull the system lower. Small incipient departures from neutral stability may be problematic to detect directly, that is, incipient departure may have to be deduced from position or velocity dispersions alone.

• Note that the balance point depends upon the mass distribution of the system, that in turn, depends upon the ribbon's density and taper, the amount of ribbon deployed, and the fuel remaining in the end-craft. The ribbon must be maintained *delicately poised* between the conflicting tendencies of centrifugal and gravitational effects with no control effectors other than (a) position state of the end-masses, (b) distributed mass within the ribbon, and (c) onboard propulsion.

• Insuring stability (for anchor grappling) will require active-propulsion since *ribbon deployment,* per se, may prove ineffective for overcoming departure from the balance point. Such imbalance can result from factors ranging from uncertainty in state-recognition (obscuring detection of an insipient departure at deploy termination), to the transport delays inherent in control inputs propagating the length of the ribbon, thus attenuating the effects of control inputs related to deploy rate modulation; note, time for tension gradients to traverse the ribbon are about 20 minutes from earth to GEO, and 45 minutes from earth to Ballast.

• In order to *minimize* the need for onboard propellant, the progression of intermediate states comprising the deployment must all be delicately balanced between gravitational attraction and centrifugal effects; this becomes increasing problematic as the Deploy craft approaches increasingly non-linear *lower* regions of the inverse square gravity field.

• As the ribbon is extended up, a tangential velocity *make-up* is required to maintain effective angular rate consistent with the earth angular velocity; failure to do this will compromise the necessary centrifugal counter-balance effect. At Ballast altitude the required tangential velocity is 7292 m/s (23,900 ft/s) relative to the anchor point (note, this is on the order of LEO insertion velocity).

- *Orders of magnitude* change in ribbon *effective end-to-end spring rate, length, and tension* are experienced over the course of this deployment. Initially end-to-end spring-rate and related natural frequencies can be quite high, but, near terminal phase (when vertical control near earth becomes critical), the ribbon will exhibit a spring rate on the order of .004 N/m. End-mass bobbing mode frequencies, and longitudinal and transverse string mode frequencies of the ribbon system change drastically over the deployment. This means that control systems must adapt to a vast range of frequencies potentially compromising control precision.

9. Deployment Phase Definitions and a Control Scenario

This is a summarization of a strategy and control scenario that has proved useful for envisioning deployment of the elevator from an initial GEO position.

Initial Phase:

This could be accomplished by ejection of the Deploy craft with a ribbon deploy-rate slightly greater than Deploy craft ejection rate. As the Deploy craft recedes into the gravity field, it slowly accelerates, removing ribbon slack; Prior to realizing tension, the Deploy craft will simply progress below and forward of the GEO craft in accordance with relative orbital motion (per Clohessy-Wiltshire equations). When the ribbon finally goes taut, tension will cause the Deploy craft will begin a harmless libration relative to the GEO craft. This libration is naturally damped, becoming inconsequential to the overall deployment. This maneuver requires virtually no control intervention by the Deploy craft (except minimal attitude control to avoid ribbon entanglement). This Initial Phase is not a critical mission phase from a dynamics standpoint. The design criteria would be to simply get some ribbon deployed and the Deploy craft sufficiently removed from the GEO craft to enable continuing gravity gradient driven separation. Tension would be kept to a minimum to facilitate the growing departure between the craft. Ideally this phase would be accomplished with minimal propulsion by both craft.

Mid Phase:

This phase will be a long duration maneuver during which the majority of the ribbon will be deployed. As deployment progresses towards consequential tension buildup, the GEO craft must take action to counter this. To avoid being pulled down, either direct equilibrating vertical thrust must be provided (with significant fuel budget consequences), or, *dynamic equilibration* of this mounting tension achieved. A method to achieve dynamic equilibration is outlined below:

- a desired Deploy craft Altitude-rate -vs- Altitude profile is <u>*indirectly*</u> commanded as an expression of (compensated) <u>*ribbon deploy rate*</u>,

in conjunction with the above deployment, the GEO craft is controlled such that,

- the GEO craft Vertical translational control algorithm attempts to achieve an altitude at which centrifugal effects *fully equilibrate* the tension and gravitational acceleration being realized at the GEO craft.

- the GEO craft Horizontal translational control algorithm provides tangential velocity make-up to ensure centrifugal equilibration effectiveness and limit libration oscillations that might adversely couple with vertical dynamic modes.

Note that for Mid Phase deployment, vertical and horizontal control may not be necessary for the **Deploy craft**. Mid Phase terminates at the atmospheric interface. By Mid Phase termination, Deploy craft altitude rate will have been stabilized and controllable via a combination of ribbon fine-deployment, and propulsive control.

Atmospheric Phase:

Atmospheric traversal may entail (a) Delaying atmospheric encounter until that time when minimum wind conditions prevail, (b) Propulsive control *closing the loop* on earth position sensing. This phase was not simulated in this paper.

Terminal Phase:

Terminal phase consists of the combined actions of *fine control* of earth position, altitude, and altitude-rate. Altitude rate control would likely be accomplished by propulsion in conjunction with vernier ribbon deployment. This phase was not simulated in this paper.

10. Dynamics of a Possible Control Algorithm for Deployment

The above described deployment mission scenario demonstrates the possibility of dynamically balancing the vertical ribbon during the course of deployment and suppressing un-desirable dynamic ribbon responses, all by means of control effectors of *significantly less force* than the *steady tensions* being managed during the deployment. This technique is an interplay of a Vertical and Horizontal controller for the GEO craft *combined with* a ribbon deployment scenario that modulates the ribbon deployment-rate as a function of the Deploy craft altitude, while paying due regard to a supplementary deployment rate component required to compensate for the rising altitude of the GEO craft itself. Ideally, the GEO craft Vertical control is of such precision as to require virtually no static propulsive-makeup against tension; in which case, a minimum propulsive impulse would roughly correspond to the sum of the *work* that must be done to vertically control the GEO craft through the gravity field from GEO to Ballast altitude (a quantity highly sensitive to optimal design) plus tangential velocity make-up (an essentially fixed quantity). This controller uses logic to minimize modal interaction with the combined ribbon/end-body system, while counteracting the intrinsic tension and gravity state; problematically, the vertical controller can induce spurious tension transients into the ribbon system in the act of maneuvering, then in turn, react to these very transients. For this reason the vertical controller uses *filtered tension sensor* data to plan maneuvers (along with other schemes to counteract instabilities). Details of the control algorithms and deployment scenario follows.

GEO craft control:

For the Horizontal axis, conventional on/off control logic was employed to maintain the GEO craft to within a specified dead-band of a specified fixed earth longitude and latitude. This control logic is combined with a Coriolis bias that commands a horizontal thrust-level proportional to the altitude rate and earth rotation rate. The maximum horizontal thrust allowed for this mode was 2200 N.

For the Vertical axis, a conventional error/error-rate feedback proportional controller commanding a maximum of 6500 N of thrust was used. This controller commanded an altitude that would be consistent with equilibrating the ribbon tension with due regard for local gravitational acceleration; this equilibration assumes a tangential velocity corresponding to the GEO craft's position *as though it were on a vertical radial rotating with the earth*.

Deploy craft control:

It was determined that due to the inherent *relative libration stability* of space tether deployment, no active translational control was needed on the Deploy craft for the *Initial-phase* and *Mid-phase* of this deployment.

The Ribbon Deployment:

Ribbon deployment rate is the sum of two contributions: (a) the baseline rate profile from a table of *Deploy-rate -vs- Deploy craft Altitude* representing a desired rate of descent for the Deploy craft, and (b) the *altitude rate* of the GEO craft (as it rises to equilibrate the ever-building tension). This algorithm is configured to inhibit negative deployment rate so as to attenuate deployer participation in longitudinal dynamic modes and GEO craft controller-induced vertical dynamics. A strain-bias of 0.075 (based on a reference tension of 22,000 N) accounts for the fact that the deployer algorithm dispenses *un-elongated* tether, which, upon being emitted into the domain of the tether, is destined to acquire a strain consistent with the level of stress extant in the tether.

Figure 14 below shows the *base-line* deploy rate for this example, and derived from an idealization of a possible *Altitude-rate -versus- Altitude* profile that might be appropriate for a Deploy craft to experience. The ribbon deploy rate is commanded as *this baseline value*, plus, the Altitude rate of the GEO craft.

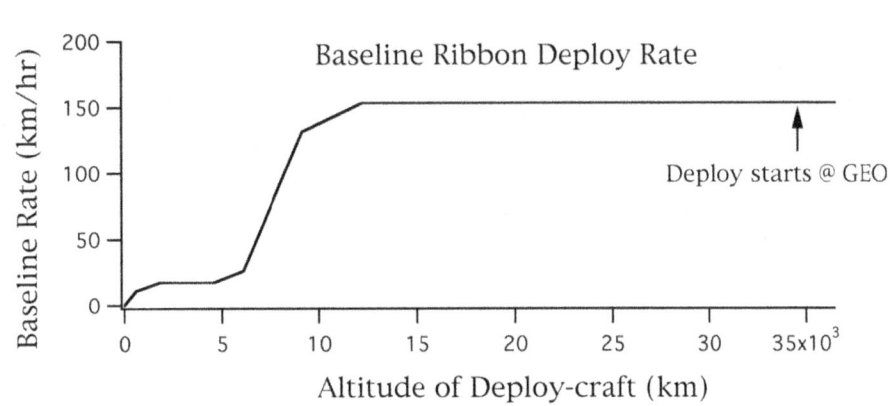

Figure 14. Baseline Deploy Rate Component -vs- Altitude of Deploy craft

Figure 15 shows the composite ribbon deploy rate experienced by this mission.

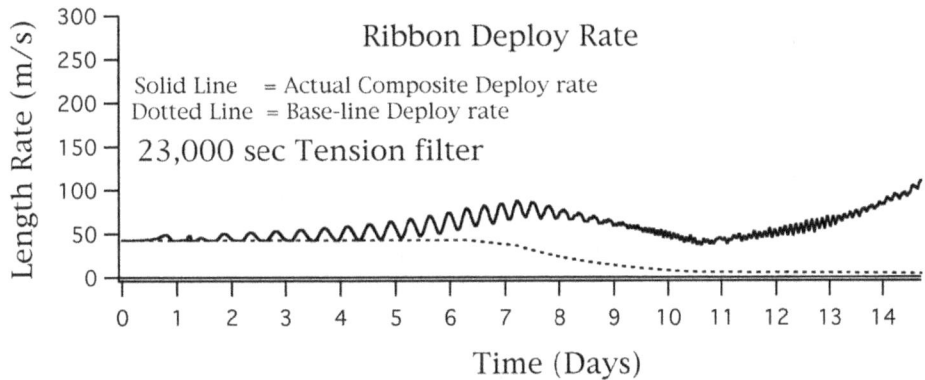

Figure 15. Final Composite Deploy Rate Reflecting GEO craft Altitude Rate

Shown in Figures 16 and 17 are the actual altitude rates achieved by the Deploy craft and GEO craft, and clearly show rate modulation starting at about 8 days.

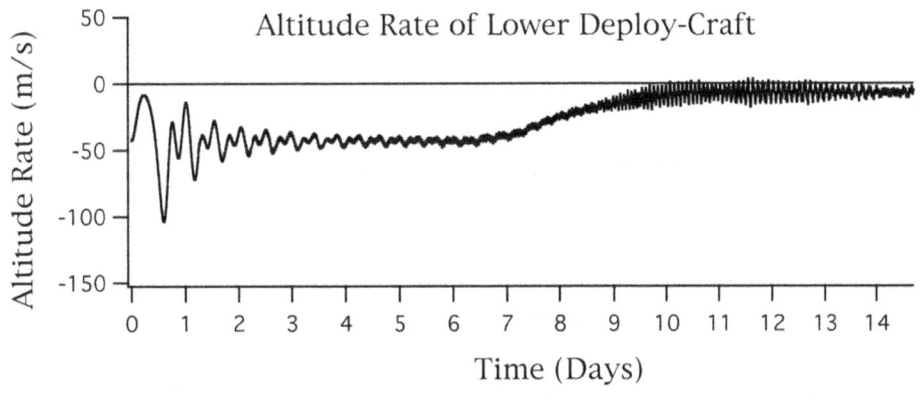

Figure 16. Altitude Rate of Deploy craft

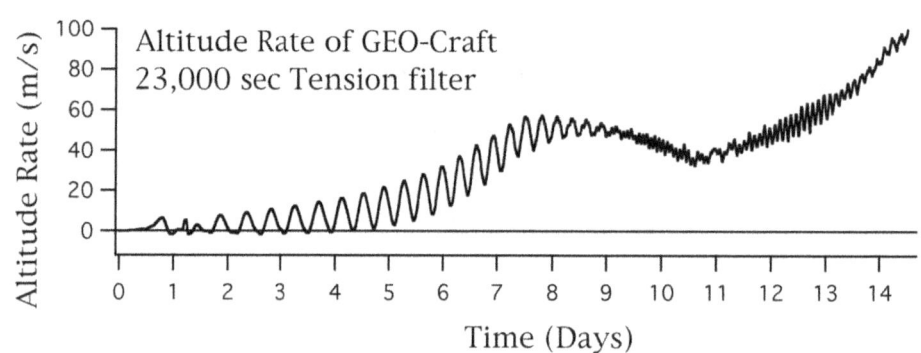

Figure 17. Altitude Rate of GEO craft

The resulting altitude profiles of the GEO and Deploy craft are seen in Figure 18.

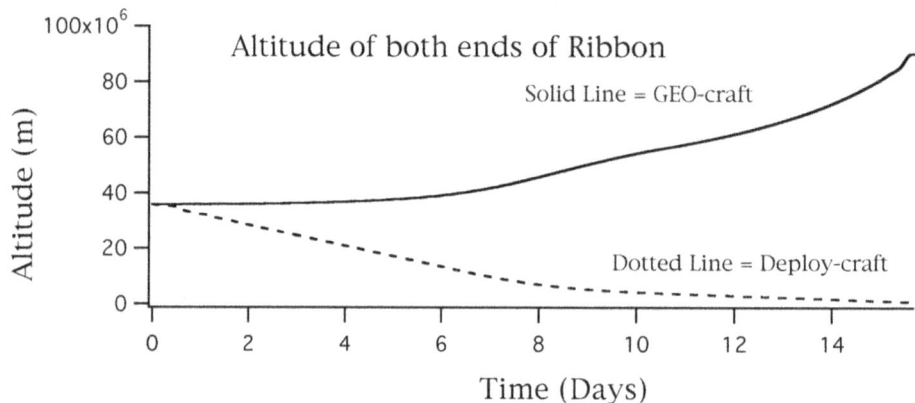

Figure 18. Altitude of GEO craft and Deploy craft

The ribbon tension is shown in Figure 19.

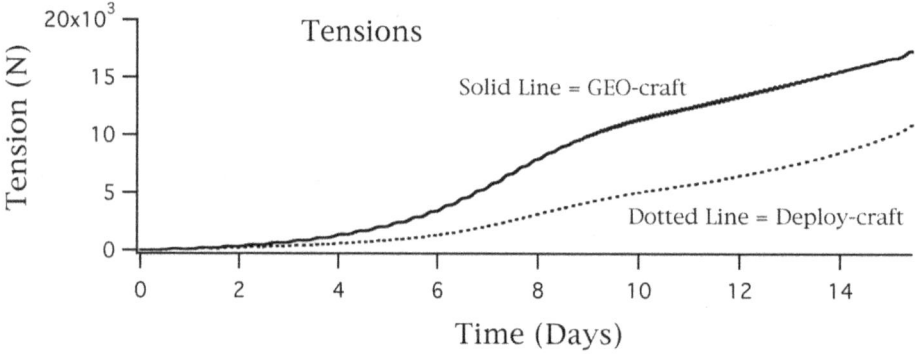

Figure 19. Tension

Libration experienced by the GEO craft is shown in Figure 20. Notice that the horizontal dead-band controller bounds the libration at about the +1 degree dead band indicating that the tangential velocity makeup control was slightly biased.

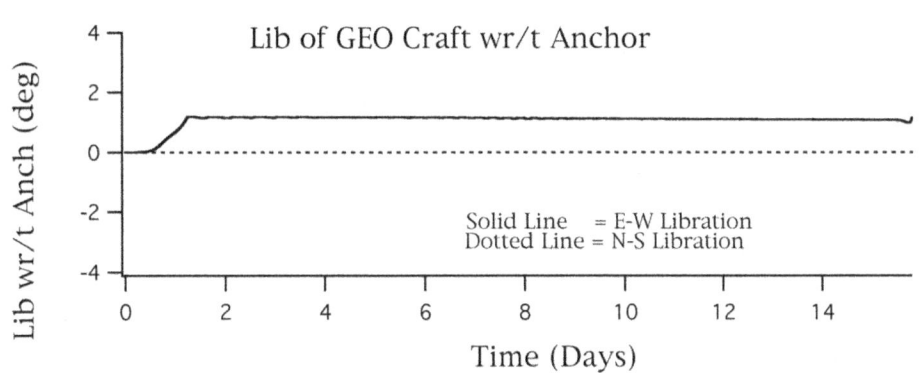

Figure 20. GEO craft Libration relative to Anchor Point

Libration experienced by the Deploy craft is shown in Figures 21. Even though the Deploy craft has no active control, its initial libration perturbation (associated with tension onset) attenuates over time, a behavior typical of tether deployment.

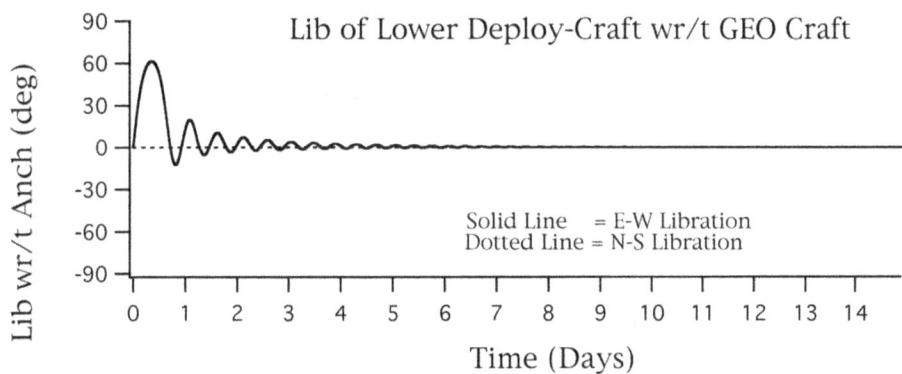

Figure 21. GEO craft Libration relative to Anchor Point

Deploy craft position relative to the anchor is shown in Figure 22. Note, if the GEO craft were initially positioned 1.5 deg. west of the anchor, the resulting eastward dead-band bias would position the Deploy craft directly over the anchor.

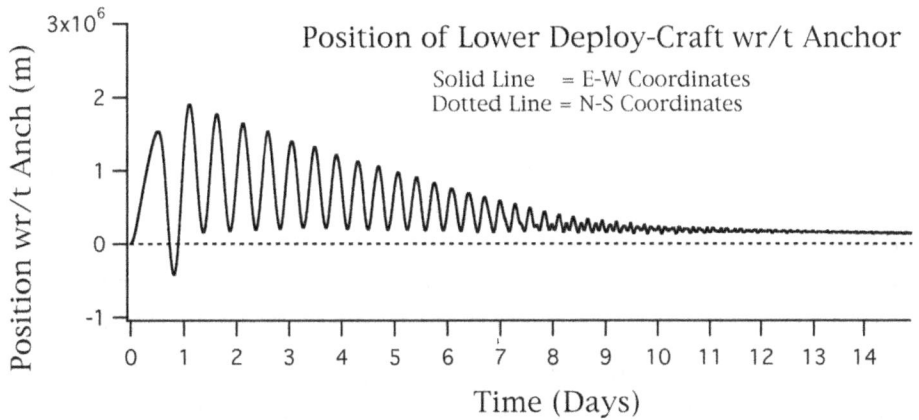

Figure 22. Deploy craft Position Error Relative to Anchor Point

11. Importance of Horizontal Control

Examination of cases in which horizontal control was not active on the GEO craft clearly indicates the critical need for tangential velocity *compensation* related to Coriolis acceleration effects. For instance, in the example above of a successful deployment, by simply disabling horizontal control on the GEO craft (but with the *successful* vertical control modes *still* enabled), the system crashes to earth. This is because the centrifugal effect is the paramount mechanism in the physics of the elevator, and as the GEO craft rises, failure to introduce tangential velocity make-up allows the GEO craft to drift into a higher "orbit" with a corresponding longer orbital period, hence, a lesser effective *angular velocity*. Thus the intrinsic centrifugal effect on which the elevator relies is removed; under these conditions the ribbon tension can easily pull-down the GEO craft.

12. Conclusions

Due to the inherently unstable attributes of a tethered system whose length spans a significant portion of the gravity field, as in the case of the space elevator, it appears that active control effectors will be necessary to perform this mission. Failure to accomplish such control was found to easily result in total loss of the initial elevator system since un-attenuated vertical imbalances result in either the entire system collapsing to earth, or flying off into an irretrievable trajectory.

While such control can be doubtlessly accomplished given sufficient propellant budget, the engineering challenge facing an actual deployment is to achieve control and stability within practical levels of total propulsive impulse expenditure. The ideal lower-bound on mission impulse could be thought of as the sum of the impulse required to achieve tangential velocities consistent with earth rotation, plus the impulse required to lift the ballast (and ribbon) against the gravity potential. Since expenditure of total impulse has mission-lapsed-time implications (analogous to gravity losses for classical rocketry), short mission durations are desirable; however, stability and safety of deployment speaks for long slow deployments, thus, the mission design will likely entail compromises related to this area of performance.

Only *insignificant* transverse ribbon oscillation modes were excited during the process of deployment. While, this was not true during the engineering development process for the various control modes and deployment strategies, it was found that as deployment scenarios started to meet mission objectives successfully, then simultaneously, transverse string mode deflections became inconsequential. This was probably because *successful* deployment schemes (almost axiomatically) manifested themselves as *smooth* deployment processes.

A proposed control law and deployment scenario has been simulated and found to demonstrate the possibility of effectively managing the space elevator ribbon deployment down to the atmospheric phase interface.

13. Future Work

Many areas of new investigation regarding initial ribbon deployment invite further exploration. The optimization of control algorithms will be critical to accomplishing the mission, yet expending an affordable total impulse. Transit though the atmosphere was not addressed in this study; the effective use of propellant in this phase of the mission may be critical. Finally the terminal phase of the mission in which contact is made with the anchor station will require sensitive control of altitude rate of the Deploy craft, as well as the development of innovative grappling schemes and maneuvers.

The potential benefits of a LEO originating deployment were not addressed in this study. The dynamic responses of such an approach should be addressed next in order to determine which mission scenario might be optimal for the initial space elevator deployment.

Acknowledgements

Funding for this work has been provided in part by the Institute for Scientific Research, Fairmont, WV and the NASA Marshall Space Flight Center.

References

1. Edwards, Bradley C., Westling, Eric A. "The Space Elevator", published by Spageo Inc, San Francisco, CA, 2002.

2. Edwards, Bradley C., unpublished communications with David Lang, 2002.

3. Lang, David D., "Space Elevator Dynamic Response to In-Transit Climbers", proceedings of the Space Engineering and Space Institute, 2005.

4. Lang, David D., "Approximating Aerodynamic Response of the Space Elevator to Lower Atmospheric Wind", proceedings of the Space Engineering and Space Institute, 2005.

5. Pearson, Jerome, "The Orbital Tower: a Spacecraft Launcher Using the Earth's Rotational Energy", Acta Astronautica. Vol. 2. pp. 785-799

6. Shelef, Ben, "LEO Based Space Elevator Ribbon Deployment", Gizmonics, Inc. Unofficially published (un-dated)

LUNAR ANCHORED SATELLITE TEST

Jerome Pearson, is, of course, the American Engineer who independently invented the idea of a tensile based Space Elevator, the model that all serious Space Elevator research follows today. He and fellow inventor, Russian Yuri Artsutanov, laid the groundwork for all that followed and that continues to happen in this field. Jerome is also the first person who put together the rigorous mathematical underpinnings describing the behavior of not only an earth-based Space Elevator, but a Lunar-based one too.

The following document is of great historical interest. As Jerome describes it, *"In 1978, I presented an AIAA paper at the Astrodynamics Conference on the "Lunar Anchored Satellite Test, which includes the mathematics that were later used in the Journal of the Astronautical Sciences in 1979"*. This is the first publication of the full paper.

We at CLIMB are very grateful for being allowed to publish this early Space Elevator-related document.

Sponsored by—
American Institute of Aeronautics and Astronautics (AIAA)
American Astronautical Society (AAS)

78-1427

Lunar Anchored Satellite Test

J. Pearson, *Flight Dynamics Lab, Wright-Patterson AFB, Ohio*

AIAA/AAS ASTRODYNAMICS CONFERENCE

Palo Alto, Calif./August 7-9, 1978

For permission to copy or republish, contact the American Institute of Aeronautics and Astronautics, 1290 Avenue of the Americas, New York, N.Y. 10019.

LUNAR ANCHORED SATELLITE TEST*

Jerome Pearson**
U. S. Air Force Flight Dynamics Laboratory
Wright-Patterson Air Force Base, Ohio

Abstract

A full-scale test is proposed for a new concept in large space structures--the anchored satellite. A satellite placed somewhat beyond the L2 libration point behind the moon would be attached to the center of the moon's farside by a tapered cable 70,000 km long. Existing high-strength composite materials could withstand the required stress with area tapers of 30 to 100. This lunar anchored satellite would follow a Lissajous path behind the moon, providing virtually continuous communication between the earth and the lunar farside. It would require no station-keeping propellant, making it potentially very long lived.

I. Introduction

This paper proposes the first full-scale test of a new kind of large space structure--the anchored satellite. As the name implies, an anchored satellite is attached to the body about which it orbits. The structure is composed of an extremely long, thin wire or cable balanced in tension about the synchronous orbit radius and extending from the surface of the planet to a counter-weight placed above the equilibrium point. The general features of an anchored satellite, or "orbital tower," are shown in Figure 1. The structure is maintained in uniform stress by an exponentially varying cross-sectional area which would reach zero at the planet's center and at infinity. The upper end can be truncated at any point above the synchronous point r_s and still be kept in uniform stress if an appropriate counterweight is added. The satellite then exerts an upward force on the planet's surface of σA_0, where σ is the uniform stress and A_0 is the base area; this force represents a net lifting capacity. Such a planet-to-orbit connection would allow ordinary electrical propulsion systems to be used for launching and retrieving space payloads, eliminating booster rockets and heat shields. By using the energy of the earth's rotation, such an earth-anchored satellite could launch payloads from synchronous orbit to the moon and other planets with no additional energy required.

The concept of a structure which extends from the ground into orbit was first proposed by Artsutanov[1] and independently by Isaacs et al.[2], and by Pearson.[3] This concept requires an exponential taper in the cross-sectional area of the

* Copyright © 1978 by Jerome Pearson, with release to AIAA to publish in all forms. This research is not part of the author's official duties.

** Aerospace Engineer, Associate Fellow AIAA.

RELEASED TO AIAA TO PUBLISH IN ALL FORMS

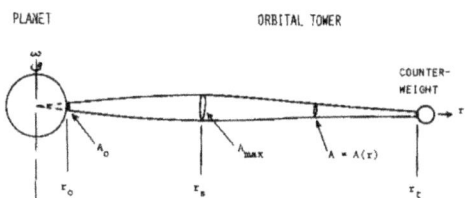

Fig. 1 General features of the anchored satellite, or orbital tower.

cable which depends on the strength-to-density ratio of the building material. For the earth-anchored satellite, unfortunately, reasonable taper ratios of 10 to 100 between the cross-sectional area at synchronous orbit and that at the ground would require strength-to-density ratios which exist only for perfect-crystal materials. There are also many dynamics problems to be overcome.[4]

The construction of such large-scale space structures is now being approached from the direction of greatly elongated satellites. Communications antennas several kilometers long have been proposed, and Chobotov[5] has analyzed a gravity-gradient stabilized solar power satellite 58 km long in earth orbit. A satellite has also been proposed which would hang from the space shuttle into the upper atmosphere on a 100-km long tether.[6] The deployment and retrieval of the shuttle-tethered satellite have been analyzed and shown to be feasible. This tethered satellite is planned for an early shuttle flight.

A second approach is to construct a lunar anchored satellite. A satellite anchored to the moon would be far less demanding of materials than the earth-anchored satellite. Isaacs et al. suggested a connection between the moon's farside and the unstable Lagrangian point L2, but did not analyze this situation. Pearson[7] first analyzed the requirements for anchored lunar satellites about L1 and L2 and found them to be within the capabilities of existing high-strength materials. The problem addressed in this paper is a specific proposal to apply the anchored lunar satellite theory to a useful space mission which could be performed within the next decade. The proposed mission is to deploy a communications satellite beyond L2 on the farside of the moon and anchor it to the lunar surface. Such an anchored relay could be used for nearly continuous communications between the earth and roving vehicles on the lunar farside. This mission would bridge the gap between the 100 km-long tethered satellite for the shuttle and the extremely difficult earth-anchored satellite.

II. The Lunar Anchored Satellite

The proposed satellite anchored about L2 is shown in Figure 2. The tapered tether plays the part of the orbital tower and the satellite is the counterweight. The variation in cross-sectional area is given by[7]

$$A(x) = A_{max} \exp\left[\frac{.0002779}{h}\left(\frac{1-\mu}{\mu-x} + \frac{\mu}{\mu-x-1} - \frac{x^2}{2} - k\right)\right] \quad (1)$$

where $k = \frac{1-\mu}{\mu-x(L2)} + \frac{\mu}{\mu-x(L2)-1} - \frac{x^2(L2)}{2}$

and x is the position along the tether, h is the characteristic height (strength-to-weight ratio) of the tether material, and μ is the ratio of the moon's mass to the total earth-moon mass, $\mu = 1/82.30$. Both x and h are normalized to the earth-moon distance and h is defined as $h = \sigma/\rho g_e$, where σ is the material tensile stress limit, ρ is the density, and g_e is earth's surface gravity. The characteristic height of a material is the maximum length without breaking of a uniform-diameter hanging wire of the material under one earth gravity. For balance, the satellite must be beyond L2.

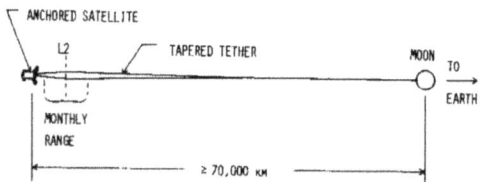

Fig. 2 Proposed lunar anchored satellite.

The taper ratios required for the anchored lunar satellite about L2 and also about L1, on the near side of the moon, are shown in Figure 3 as functions of the characteristic height of the material. The taper ratio is defined as the ratio of the cross-sectional area at the balance point to the cross-sectional area at the base. This figure shows the requirement for the earth-anchored satellite for comparison. The L1 and L2 anchored lunar satellites are seen to be nearly identical in requirements and to be far less demanding of material strength than the earth-anchored satellite. They could be constructed from existing engineering materials[8] such as those shown in Table 1. For example, the graphite/epoxy composite could be used to construct anchored lunar satellites with a taper ratio of 30. This value will be assumed for succeeding calculations.

Table 1 Composite materials properties

	DESIGN STRENGTH GN/m²	DENSITY kg/m³	CHARACTERISTIC HEIGHT, km
GRAPHITE/EPOXY	1.24	1550	81.6
BORON/EPOXY	1.32	2007	67.3
KEVLAR 49	0.703	1356	52.9
BORON/ALUMINUM	1.10	2713	41.5

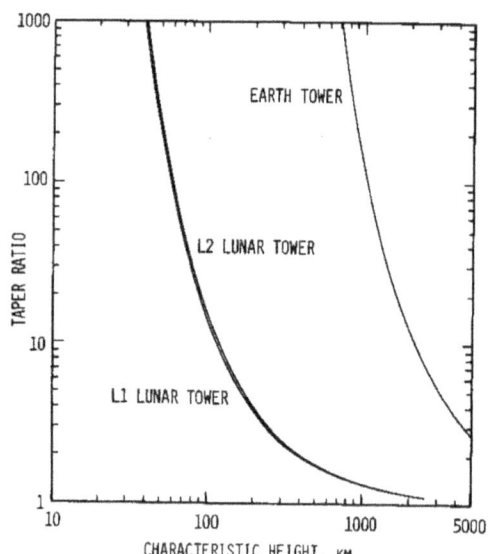

Fig. 3 Taper ratios required for anchored satellites.

The amount of material required to build the L2 tower is shown in Figure 4 in terms of the tower mass and counterweight mass per unit base area. These curves are based on a taper ratio of 30, corresponding to h = 80.95 km, very close to the value for graphite/epoxy. The tower mass is shown dashed below L2, the minimum height for balance. A counterweight (satellite) placed just above the L2 balance point would need to be very massive to support the weight of the tower below. At higher locations the counterweight feels a greater gravitational acceleration upward, so less mass is required to balance the tower. By properly choosing the location of the satellite, the desired ratio of satellite mass to tether mass may be selected.

The variation in cross-sectional area with distance along the tether is shown in Figure 5, normalized to the base area. This diagram is based on a selected satellite distance of 70,000 km from the moon's center. This allows an adequate margin of safety for the monthly variation in the location of L2 from 60,974 to 68,057 km from the moon's center.

The stability of a body on a massless tether in the three-body system of earth-moon-spacecraft, assuming a circular lunar orbit, has been investigated[7] and the results are summarized in Figure 6. In a rotating coordinate system with the earth at μ and the moon at $\mu-1$, the stability boundaries for L1 and L2 satellites are seen to be at right angles to the tether length. In this figure the locations of the Lagrangian points L1, L2, and L5 are shown to scale, along with values of the Jacobian constant C. The stability boundaries are shown as dashed lines; the "s" side is stable and the "u" side is unstable. A satellite located 70,000 km behind the moon's center could thus

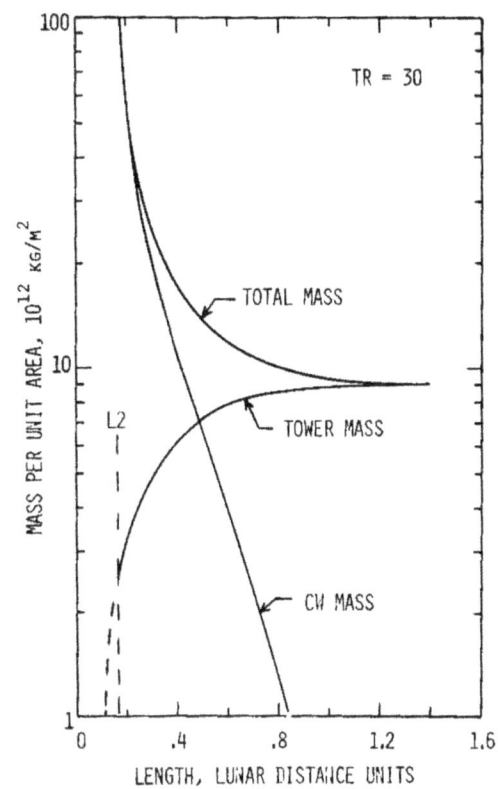

Fig. 4 Mass and counterweight requirements for the lunar anchored satellite.

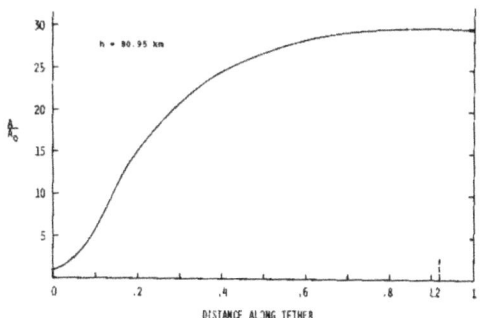

Fig. 5 Variation in cross-sectional area for the lunar anchored satellite.

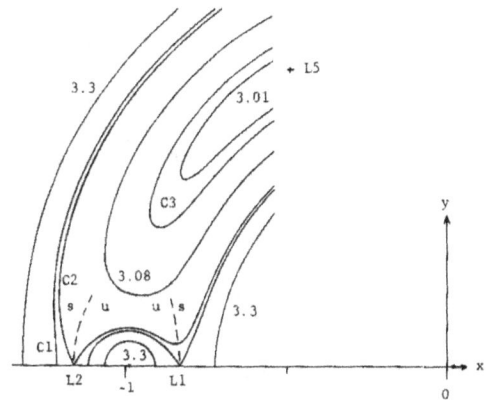

Fig. 6 Stability boundaries for L1 and L2 lunar anchored satellites.

experience excursions of as much as 30 degrees on a taut tether and remain stable.

The period of this spherical pendulum in the z direction is different from that in the y direction (in the plane of the moon's orbit) and both are functions of the amplitude of motion. A typical path is shown in Figure 7 for a period of 65 days. The path is seen to be a complex Lissajous pattern which results in occasional occultations of the satellite by the moon as seen from L1 (shaded area), or from the earth. The proportion of lost communication time is about equal to the projected area of the moon divided by the area of the path bounds. For the maximum excursion of ± 30° in the x-y plane the lost time is about 0.2%, and for a 3° excursion the lost time is about 28%, for communication with the earth. There is thus a clear advantage to larger paths.

III. Construction Technique

To emplace the L2 anchored satellite it is necessary to extend the tether toward the lunar surface from a stabilized satellite very near L2. The technique described here can be carried out with existing propulsion systems. For example, the Titan/Centaur combination is capable of placing 3900 kg of mass at L2. For a satellite with a final location at 70,000 km behind the moon, the required masses from Figure 4 would be 3778 kg for the counterweight and 122 kg for the tether. For such a small tether mass, the minimum cross-sectional area of the base becomes very small: $A_0 = 4.20 \times 10^{-11}$ m^2. This area would allow only a few individual graphite fibers to be used, increasing the risk of breakage.

This problem can be alleviated in several ways. First, the satellite can be placed at a higher altitude; Table 2 shows the change in minimum tether area for distances up to 90,000 km from the moon's center. Changing the distance to 80,000 km nearly doubles the allowable base area for a given total weight, and changing the distance from 80,000 to 90,000 km increases the base area another 24%. Second, the total amount of inert ballast carried by rocket to L2 could be increased, depending on the payload capacity of the launch vehicle. This larger counterweight would then require a larger tether to balance it. The third and perhaps best method to increase the number of base fibers is to use different material for the smallest-diameter part of the tether.

Table 2
Lunar anchored satellite parameters vs location

r, km	70,000	80,000	90,000
Total mass, kg	3900	3900	3900
satellite mass	3778	3631	3528
tether mass	122	269	372
$A_o, 10^{-11} m^2$	4.20	7.89	9.80

Fibers of crysotile, for example, are so small[9] that thousands of them could be packed into the smallest base area of Table 2.

The construction technique is then as follows. A Titan/Centaur vehicle is launched from the Eastern Test Range and the Centaur guidance and restart capability is used to bring the payload to L2. Experience for this maneuver will have been gained from the launch in 1978 of the International Sun-Earth Explorer (ISEE) satellite into a halo orbit about the L1 point of the sun-earth system.[10] Using the ISEE technique, the Centaur can be placed into a halo orbit about L2 which has an in-plane radius of at least 2000 km to ensure constant visibility from the earth during the tether attachment. An unreeling mechanism and a drum of the tether line packaged in the Centaur will be deployed next, using techniques developed for the Shuttle-tethered satellite.[11] It will be necessary for the end of the cable to include a small transmitter for tracking purposes during the tether unreeling. A small, cold-gas thruster will also be needed to start the tether deployment and to take the tether end several kilometers below the Centaur in order to develop a tension in the line by the gravity gradient. After a large enough tension is developed, a passive unreeling of the tether will suffice. An alternate procedure is to unreel several hundred meters of line and then forcibly eject the tether end and transmitter downward.[12] After a few bounces, the line will become taut. The tether line will then follow the gravity gradient to the moon's surface near the center of the farside. The lengthening of the line adds damping and stabilizes the pendulum motion.[13] The tension and the tether end position can be monitored to control the rate of deployment; at an average deployment rate of 20 meters per second, several weeks will be required before touchdown. This average rate corresponds to 130 revolutions per minute for a large reel inside the 3-meter diameter of the Centaur. The acceleration and deceleration of this reel will require additional attitude-control impulses from the satellite stationkeeping system in addition to those required to maintain the unstable halo orbit.

Once the tether has touched down, a penetrometer or an inert mass can anchor the line and then the satellite can back off from L2 to its final distance of 70,000 to 90,000 km. At this point, several weeks into the mission, the stationkeeping and attitude control system can be turned off. The earth and moon both can be maintained in the beam of a communications antenna with a beam-width of a few degrees without tracking. Without a need for further stationkeeping, the anchored lunar satellite could

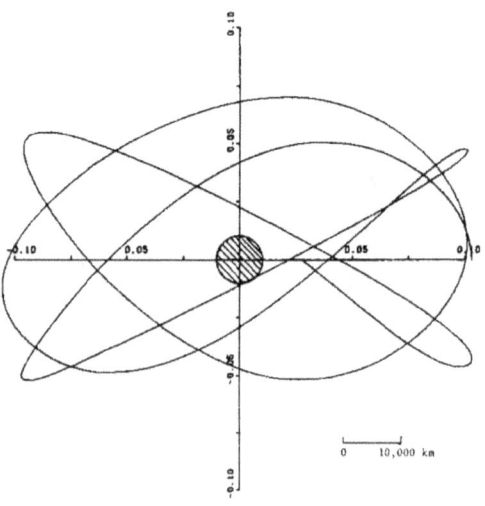

Fig. 7 Path of L2 lunar anchored satellite as seen from L1 (moon-occulted area shaded).

provide a longer-term lunar farside communication system than the proposed lunar halo satellite.[14]

IV. Concluding Remarks

There are several additional studies which are necessary before this mission can be undertaken. In particular, dynamics studies of the deployment of extremely long tethers are needed. No study to date has dealt with deployment dynamics over distances comparable to the satellite orbital radius. The behavior of graphite/epoxy and chrysotile fibers under stress in the space environment for a period of years must be known before the final choice of tether material can be made. Finally, the radiation and meteoroid environments near L2 need better definition. The effect of a collision on an anchored satellite could be catastrophic. Also for this reason, the number of man-made objects in lunar orbit should be severely limited, and where possible they should be put into orbits which will decay rapidly after their missions are accomplished. The anchored lunar satellite could provide a more detailed knowledge of L2 dynamics and could become the basis for a larger installation used to develop lunar resources. For these reasons the lunar anchored satellite test is an exciting prospect for future lunar development.

V. References

1. Artsutanov, Y., "V kosmos na electrovoze," Kmosmolskaya Pravda, 31 July 1960. (A discussion by Lvov in English is given in Science, Vol. 158, pp. 946-7, 17 November 1967).

2. Isaacs, J. D., Vine, A. C., Bradner, H., and Bachus, G. E., "Satellite elongation into a true 'sky-hook'," Science, Vol. 151, pp. 682-3, 11 February 1966.   (See also Science,

Vol. 152, p. 800, 6 May 1966).

3. Pearson, J., "The orbital tower: a spacecraft launcher using the earth's rotational energy," Acta Astronautica, Vol. 2, Sep/Oct 1975, pp. 785-99.

4. Pearson, J., "Using the orbital tower to launch earth-escape payloads daily," presented at the 27th IAF Congress, Anaheim, Calif., 10-16 October 1976. AIAA paper IAF 76-123.

5. Chobotov, V. A., "Gravitationally stabilized satellite solar power station in orbit," Journal of Spacecraft and Rockets, Vol. 15, No. 4, pp. 249-51 April, 1977.

6. Weiffenbach, G. C., Colombo, G., Gaposchkin, E. M., and Grossi, M. D., "Gravity gradient measurements in the vicinity of 100 km height by long-tethered satellites," presented at the 27th IAF Congress, Anaheim, Calif., 10-16 October 1976. AIAA paper IAF-76-064.

7. Pearson, J., "Anchored lunar satellites for cis-lunar transportation and communication," presented at the European Conference on Space Settlements and Space Industries, London, England, 20 September 1977.

8. Anon., Advanced Composites design guide, 3rd Ed., U. S. Air Force Materials Laboratory, Wright-Patterson AFB, Ohio, 1973.

9. Parratt, N. J., Fibre-reinforced materials technology, Van Nostrand Reinhold Company, London, 1972, p. 91.

10. Farquhar, R. W., Muhonen, D. P., and Richardson, D. L., "Mission design for a halo orbiter of the earth," Journal of Spacecraft and Rockets, Vol. 14, No. 3, pp. 170-7, March 1977.

11. Baker, W. P. et al., "Tethered subsatellite study," NASA TM X-73314, Marshall Space Flight Center, Alabama, March 1976.

12. Kane, T. R., and Levinson, D. A., "Deployment of a cable-supported payload from an orbiting spacecraft," Journal of Spacecraft and Rockets, Vol. 14, No. 7, pp. 409-13, July 1977.

13. Watkins, T. C. et al., "Stabilization of externally slung helicopter loads," U. S. Army Air Mobility Research and Development Laboratory Technical Report TR-74-42, August 1974.

14. Farquhar, R. W., "The utilization of halo orbits in advanced lunar operations," NASA TN D-6365, National Aeronautics and Space Administration, Washington, D. C., July 1971.

ASTEROID SLINGSHOT EXPRESS

Ben Shelef has long been associated with the Space Elevator effort. His Spaceward Foundation was the organizer and host of the Space Elevator Games, competitions with $2Million in prizes provided by NASA as part of their Centennial Challenges program. These competitions greatly advanced the state of the art in Power Beaming.

He has written several important Space Elevator-related papers, two of which we have been privileged to carry in C^LIMB. Vol 1 / No 1 of C^LIMB included *The Space Elevator Feasibility Condition* and this issue of C^LIMB includes his article entitled *Space Elevator Power System Analysis and Optimization*. Both of the articles greatly advance our understanding of a tensile-based Space Elevator structure.

He has also written a short article on building a space elevator on an Asteroid and using that to "slingshot" a payload back to earth. We are very pleased to bring you that article here.

ASTEROID SLINGSHOT EXPRESS - TETHER-BASED SAMPLE RETURN

Ben Shelef

The Spaceward Foundation

Abstract: This paper examines the possibility of returning payloads from a spinning asteroid using a tether system and evaluates its merit in comparison to a conventional rocket-based return.

Concept

This paper examines the feasibility of returning payloads from an asteroid using a long sling powered by the rotation of the asteroid, as illustrated in **Figure 1**.

In this paper we approximate an asteroid as a body with negligible gravity, so its synchronous orbit is practically on its surface. Real asteroids have a small measure of gravity, but for practical purposes, this approximation is good enough as long as the asteroid's escape velocity is much smaller than the return delta-V.

On such an asteroid, there is no minimum length requirement for a rotating tether to stay erect – any outward-pointing tether with a mass at its end will remain taut, and the system will be in stable equilibrium. Deployment of the tether can therefore be done "bottom-up", starting with a spool on the surface and simply shooting the deploying mass out using a spring.

To launch a mass from the asteroid, we have to get it to the tip of the tether, and release it at the correct time. The mass will have a velocity of ω·r (see definitions below), and will initially travel in the plane of rotation of the asteroid, perpendicular to the tether. This technique therefore works best for asteroids with an equatorial plane that intersects the Earth.

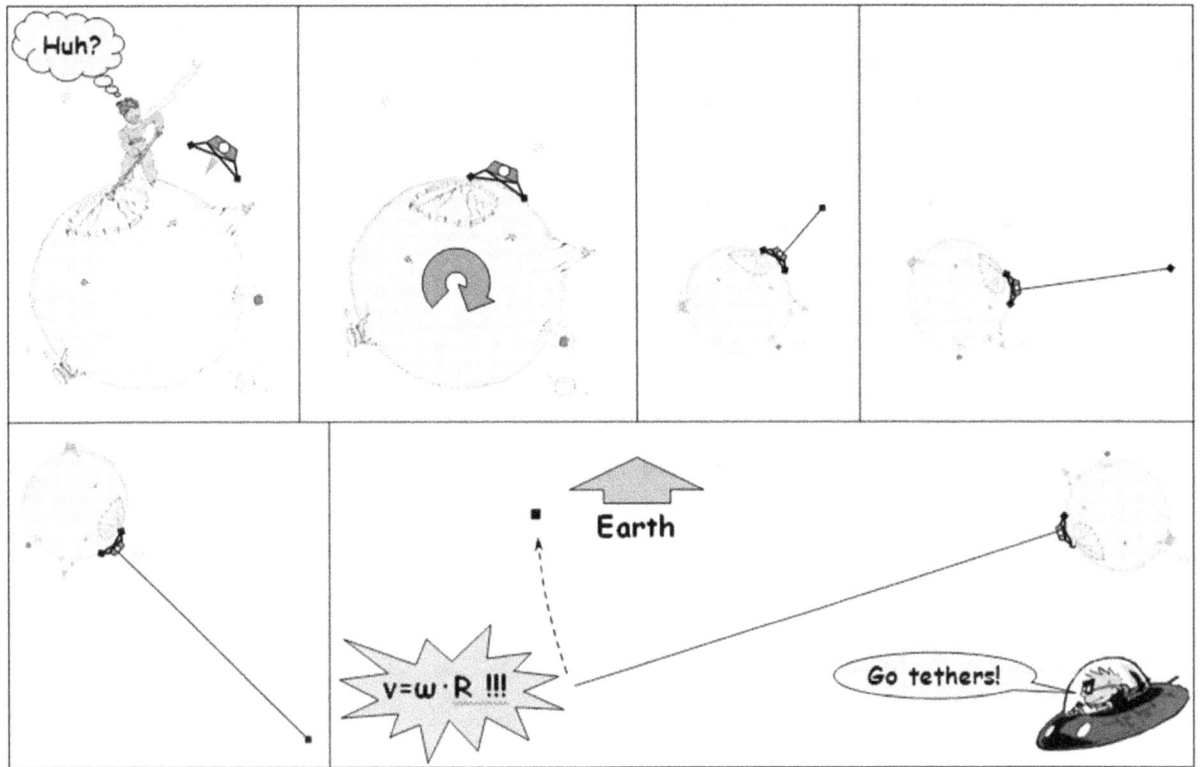

Figure 1: Asteroid Sling

When considering a single sample return, the bottom-up deployment of the tether and the launch of the sample are one and the same – there's no need to climb a pre-deployed tether. For multiple returns, once the tether is deployed, the masses will climb "down" the cable until they reach the tip. Since all motions happen with the direction of the local effective gravity, no power system of any kind is necessary. The power comes at the expense of the spin of the asteroid.

This analysis is valid only for quasi-static systems. Care must be taken during motion not to deviate too far from equilibrium, or we may risk wrapping the line around the asteroid. This is particularly true during the first deployment phase.

Tether system analysis

We will assume a small asteroid with very low gravity relative to its rate of spin such that with:

ω: Asteroid rotation rate [1/s]
r: Asteroid surface radius [m]
g: Asteroid surface gravity [m/s2]

"GEO" is near the surface
$$\omega^2 r \approx g \tag{1}$$

In which case the behavior of a tether attached to the asteroid is governed predominantly by the centrifugal acceleration, so the analysis becomes a special case of the general Space Elevator tether equation, with a diminished gravitational field.

Further denoting:

m: Payload mass [kg]

τ: Tether specific strength [μN/Tex, pa-m3/kg, or (m/s)2]

R: Tether tip distance from center of asteroid [m]

x: Distance from center of asteroid [m]

T(x): Tension of tether at x [N]

β(x): Tether linear mass density at x [kg/m]

mT: Total mass of the tether [kg]

We have the following constraints:

(Constant loading) $$T(x) = \tau \cdot \beta(x) \tag{2}$$

(Tether tension continuity) $$dT(x) = -(\omega^2 x) \cdot \beta(x) \cdot dx \tag{3}$$

(Initial condition) $$T(R) = (\omega^2 R) \cdot m \tag{4}$$

Note that unlike the Space Elevator, the initial condition in this system is at the tip, where the tip velocity is predetermined.

And so:

$$\tau \cdot d\beta(x) = -(\omega^2 x) \cdot \beta(x) \cdot dx$$
$$\beta'(x) = -(\omega^2 / \tau) \cdot x \cdot \beta(x)$$
$$\beta(x) = k_1 e^{-(\omega^2/2\tau)x^2} = k_1 e^{-(\omega x/\sqrt{2\tau})^2}$$

For convenience, we'll define the "aspect ratios" ψT of the tether system as

(Velocity characterization of a tether) $$\psi = \omega R / \sqrt{2\tau} \tag{5}$$

And rewrite:

(Linear Density at x - general) $$\beta(x) = k_1 e^{-(\psi \cdot x/R)^2} \tag{6}$$

And using the initial condition:

(Linear Density at x - specific)
$$\beta(R) = 2\psi^2 \cdot m/R$$
$$k_1 = 2\psi^2 \cdot (m/R) \cdot e^{\psi^2}$$
(7)

The area taper ratio is:

(Taper Ratio)
$$TR = \beta(r)/\beta(R) = e^{\psi^2\left(1-\left(\frac{r}{R}\right)^2\right)} \approx e^{\psi^2}$$
(8)

And the total line mass is

$$m_T = \int_r^R \beta(x)dx = \int_r^R k_1 e^{-(\psi \cdot x/R)^2} dx = k_1(R/\psi) \cdot \tfrac{1}{2}\sqrt{\pi} \cdot erf(\psi \cdot x/R)\Big|_r^R =$$
$$... = 2\psi^2 \cdot (m/R) \cdot e^{\psi^2} \cdot (R/\psi) \cdot \tfrac{1}{2}\sqrt{\pi} \cdot erf(\psi \cdot x/R)\Big|_r^R =$$
$$... = \psi \cdot m \cdot e^{\psi^2} \cdot \sqrt{\pi} \cdot erf(\psi \cdot x/R)\Big|_r^R$$
(9)

And the mass ratio (ignoring the r term)

(The tether equation)
$$m_T/m \approx \sqrt{\pi} \cdot e^{\psi^2} \cdot \psi \cdot erf(\psi)$$
(10)

Note that for a given delta-v, we can equally choose a fast rotating asteroid and s short tether, or a slow rotating asteroid and a long tether – the mass fraction is a function only of the required delta-V and the specific strength of the tether.

Specific Strength and Velocity Squared

The merit of the tether return system depends on the relative size of two velocity-united quantities: The required delta-V (ω·R) and the tether tenacity (√τ.)

Contrast this with the rocket equation, where the merit of the system is determined by the ratio of the delta-v and the specific impulse, also in units of velocity.

In other tether systems around celestial bodies, the entire effect of the body can be captured by a single velocity-unit quantity.

For example, a constant cross-section Space Elevator:

$$C_{planet} = gr_e - \frac{3}{2}\left(\omega gr_e^2\right)^{2/3} + \frac{1}{2}\omega^2 r_e^2$$, and it is feasible only if τ > Cplanet,

And for a standard Space Elevator:

$$C_{planet} = \left(\tfrac{1}{2}\omega^2 r_e^2 - \tfrac{1}{2}\left(\omega gr_e^2\right)^{2/3} + gr_e - \left(g^2\omega^2 r_e^4\right)^{1/3}\right)$$, and the taper ratio is $e^{(C_{planet}/\tau)}$

It therefore seems that the fact that the units of specific strength come out to be those of velocity is not a triviality – the specific strength tells us how much delta-v we can achieve while using the material as a sling, and which celestial objects we can construct various slings on.

Results

As an example, for an asteroid that is spinning once every 2 hours ($\omega=8.7E-4$ rad/sec), and for a 1000 m/s tether system, the length of the tether is approximately 1100 km, and the tip centrifugal acceleration is 0.87 m/sec2. A payload capsule weighing 100 kg will require a 40 kg tether, and will pull on the tip of the tether with a force of 87 Newtons (18.8 pounds). The tether cross section at the tip is therefore approximately 0.029 mm2, or 0.2 mm in diameter.

The merit of a tether system is in saving mass in comparison to a rocket-based system. A good figure of merit is the "equivalent ISP" – the ISP of a hypothetical rocket system that would weigh the same as the tether system. I assume that the rocket system has zero empty weight, which is of course conservative. Chemical rockets have ISPs between 2.5 and 3.5 km/s.

Table 2 shows several examples of tether systems using existing (green, columns A-E) and futuristic (blue, columns F-H) tethers, and their equivalent ISP. For multiple samples, the weight of the tether is divided by the number of samples.

		A	B	C	D	E	F	G	H
delta-v	m/s	500	1000	2000	5000	10000	1000	2000	5000
τ	$(m/s)^2$	3E6	3E6	3E6	3E6	3E6	10E6	10E6	10E6
ψ		0.20	0.41	0.82	2.04	4.08	0.22	0.45	1.12
TR		1.04	1.2	1.9	64.5	17E6	1.1	1.2	3.5
m_T/m		0.09	0.4	2.1	232.5	125E6	0.1	0.5	6.1
# of samples		\multicolumn{8}{c}{Equivalent ISP of tether system [km/sec]}
1		**6.1**	**3.2**	**1.8**	0.92	0.54	**10**	**5.3**	**2.5**
2		**12**	**5.8**	**2.8**	1.0	0.56	**20**	**10**	**3.6**
5		**29**	**14**	**5.7**	**1.3**	0.59	**49**	**23**	**6.2**
10		**59**	**27**	**10**	**1.6**	0.61	**97**	**45**	**10**
100		**584**	**269**	**95**	**4.2**	0.71	**968**	**438**	**84**

Table 2: The merit of the Asteroid Sling as a Function of Delta-V and Specific Strength
(Bold numbers represent cases where a tether-based system can be advantageous)

Conclusion

With NEOs and main asteroid belt objects requiring 3500-7000 m/s in delta-v, especially for multiple return masses (as in the case of asteroid mining), tether systems can quickly become more mass efficient than rocket based system. Even for the case of a single sample high delta-V return, it is still advantageous to use the sling system for the first 0.5 km/sec (in the green case) or 2 km/sec (in the blue case).

Additionally, even before Space Elevators become feasible, a tethered asteroid sample return system will be an interesting proof of concept mission.

ISEC THEME POSTERS

Each year, as part of ISEC's theme activities, ISEC produces a poster reflecting that year's theme. These posters are shown on the following pages.

- For 2009, as this was ISEC's first year, there was no theme. Instead, we chose to commemorate the Space Elevator Games (organized by the Spaceward foundation – http://www.spaceward.org with prize funding provided by NASA - www.nasa.gov/challenges/) as 2009 was the first year that a winner was awarded prize money. LaserMotive (www.lasermotive.com), out of Seattle, Washington, won $900,000 for their entry's performance. The Kansas City Space Pirates and the USST (University of Saskatchewan Space Design Team) teams are also pictured.

- For 2010, the theme was SPACE DEBRIS MITIGATION – SPACE ELEVATOR SURVIVABILITY. The poster shows two pictures. The topmost one shows a Repair Climber travelling up/down the tether, making small repairs as it detects holes in the tether caused by small pieces of space debris. The bottom picture shows a Climber using its onboard engine to induce an oscillation in the tether – this to move it out of the way of satellites and larger space debris. A combination of these techniques, plus periodic replacement of tether segments, should keep the ribbon robust and able to carry cargo.

- For 2011, the theme was RESEARCH AND THOUGHT TARGETED TOWARDS THE GOAL OF A 30 MYURI TETHER. A Yuri (named in honor of Yuri Artsutanov) is a unit of Specific Strength. It is equivalent to 1 Pascal-cubic-meter per kilogram. A Mega Yuri (MYuri) is equivalent to the commonly used units of 1 GigaPascal-cubic-centimeter per gram (1 GPa-cc/g) and to 1 Newton per Tex (N/Tex). Current thinking has a tether with a strength of 30 MYuri as being strong enough to build an earth-based space elevator.

- For 2012, the theme is OPERATING AND MAINTAINING A SPACE ELEVATOR. The poster shows several views of a proposed Space Elevator Base Station and a possible Tether Climber design.

These posters are 11x17 inches in size, are in full color, and are offset-printed on heavy-duty, glossy stock paper.

ISEC Members automatically receive that year's poster for each year they are a member in good standing. If you are not a member, and wish to purchase any of these posters, please visit the store on our website http://www.isec.org.

All of these posters were designed by Mr. Frank Chase and we at ISEC are very grateful for his artistic contributions towards our goal of building a Space Elevator. Frank can be reached at kahuna.frank@gmail.com.

ISEC 2009 THEME POSTER

ISEC 2010 THEME POSTER

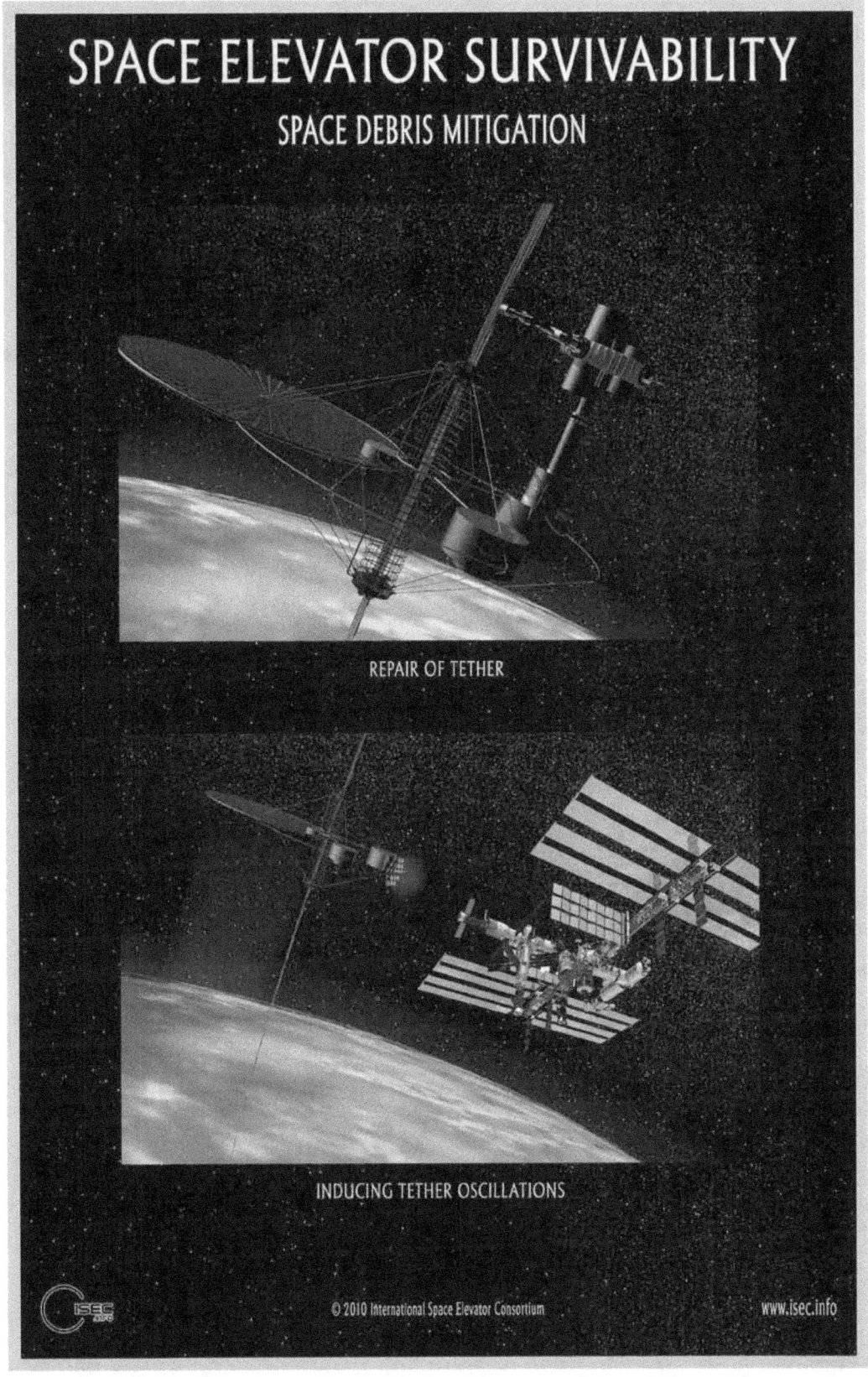

ISEC 2011 THEME POSTER

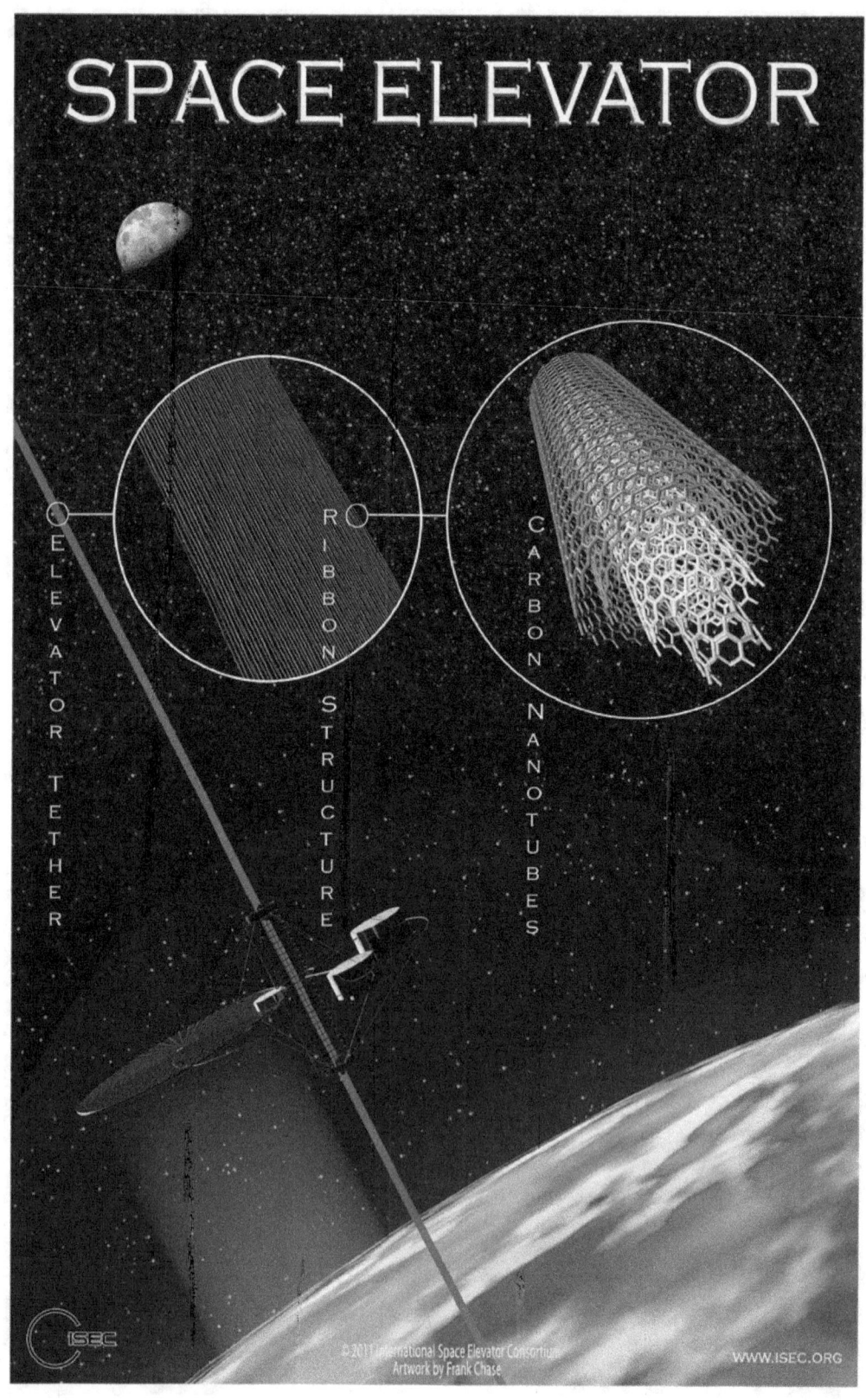

ISEC 2012 THEME POSTER

www.ingramcontent.com/pod-product-compliance
Lightning Source LLC
Chambersburg PA
CBHW080919170526
45158CB00008B/2164